Interpreting Science at Museums and Historic Sites

AMERICAN ASSOCIATION *for* STATE *and* LOCAL HISTORY

INTERPRETING HISTORY

ABOUT THE ORGANIZATION

The American Association for State and Local History (AASLH) is a national history membership association headquartered in Nashville, Tennessee, that provides leadership and support for its members who preserve and interpret state and local history in order to make the past more meaningful to all people. AASLH members are leaders in preserving, researching, and interpreting traces of the American past to connect the people, thoughts, and events of yesterday with the creative memories and abiding concerns of people, communities, and our nation today. In addition to sponsorship of this book series, AASLH publishes *History News* magazine, a newsletter, technical leaflets and reports, and other materials; confers prizes and awards in recognition of outstanding achievement in the field; supports a broad education program and other activities designed to help members work more effectively; and advocates on behalf of the discipline of history. To join AASLH, go to www.aaslh.org or contact Membership Services, AASLH, 2021 21st Ave. South, Suite 320, Nashville, TN 37212.

ABOUT THE SERIES

The American Association for State and Local History publishes the *Interpreting History* series in order to provide expert, in-depth guidance in interpretation for history professionals at museums and historic sites. The books are intended to help practitioners expand their interpretation to be more inclusive of the range of American history.

Books in this series help readers:

- quickly learn about the questions surrounding a specific topic,
- introduce them to the challenges of interpreting this part of history, and
- highlight best practice examples of how interpretation has been done by different organizations.

They enable institutions to place their interpretative efforts into a larger context, despite each having a specific and often localized mission. These books serve as quick references to practical considerations, further research, and historical information.

TITLES IN THE SERIES

1. *Interpreting Native American History and Culture at Museums and Historic Sites* by Raney Bench
2. *Interpreting the Prohibition Era at Museums and Historic Sites* by Jason D. Lantzer
3. *Interpreting African American History and Culture at Museums and Historic Sites* by Max van Balgooy
4. *Interpreting LGBT History at Museums and Historic Sites* by Susan Ferentinos
5. *Interpreting Slavery at Museums and Historic Sites* by Kristin L. Gallas and James DeWolf Perry
6. *Interpreting Food at Museums and Historic Sites* by Michelle Moon
7. *Interpreting Difficult History at Museums and Historic Sites* by Julia Rose
8. *Interpreting American Military History at Museums and Historic Sites* by Marc K. Blackburn
9. *Interpreting Naval History at Museums and Historic Sites* by Benjamin J. Hruska
10. *Interpreting Anniversaries and Milestones at Museums and Historic Sites* by Kimberly A. Kenney
11. *Interpreting American Jewish History at Museums and Historic Sites* by Avi Y. Decter
12. *Interpreting Agriculture at Museums and Historic Sites* by Debra A. Reid
13. *Interpreting Maritime History at Museums and Historic Sites* by Joel Stone
14. *Interpreting the Civil War at Museums and Historic Sites* edited by Kevin M. Levin
15. *Interpreting Immigration at Museums and Historic Sites* edited by Dina A. Bailey
16. *Interpreting Religion at Museums and Historic Sites* edited by Gretchen Buggeln and Barbara Franco

Interpreting Science at Museums and Historic Sites

Edited by Debra A. Reid,
Karen-Beth G. Scholthof,
and David D. Vail

ROWMAN & LITTLEFIELD
Lanham • Boulder • New York • London

Published by Rowman & Littlefield
An imprint of The Rowman & Littlefield Publishing Group, Inc.
4501 Forbes Boulevard, Suite 200, Lanham, Maryland 20706
www.rowman.com

86-90 Paul Street, London EC2A 4NE

British Library Cataloguing in Publication Information Available

Library of Congress Cataloging-in-Publication Data

Names: Reid, Debra A., 1960- editor. | Scholthof, Karen-Beth G., 1959-
    editor. | Vail, David D., 1981- editor.
Title: Interpreting science at museums and historic sites / edited by Debra
    A. Reid, Karen-Beth G. Scholthof, and David D. Vail.
Description: Lanham : Lexington, 2023. | Series: American Association for State and
    Local History ; 22 | Includes bibliographical references and index. | Summary:
    "Interpreting Science in Museums and Historic Sites stresses the
    untapped potential of historical artifacts to inform our understanding
    of scientific topics. It argues that science gains ground when
    contextualized in museums and historic sites"— Provided by publisher.
Identifiers: LCCN 2023021721 (print) | LCCN 2023021722 (ebook) | ISBN
    9781538172742 (cloth) | ISBN 9781538172759 (paperback) | ISBN
    9781538172766 (ebook)
Subjects: LCSH: Museums—Educational aspects. | Historic
    sites—Interpretive programs. | Science—Exhibitions. | Technological
    innovations—Exhibitions.
Classification: LCC AM7 .I587 2023 (print) | LCC AM7 (ebook) | DDC
    507.4—dc23/eng/20230602
LC record available at https://lccn.loc.gov/2023021721
LC ebook record available at https://lccn.loc.gov/2023021722

♾™ The paper used in this publication meets the minimum requirements of American National Standard for Information Sciences—Permanence of Paper for Printed Library Materials, ANSI/NISO Z39.48-1992.

Contents

Foreword

Jill Tiefenthaler

I grew up on a farm in Iowa, and I still remember the excitement of visiting museums across the state on weekends and summer days. I remember staring into museum display cases to get a closer look at the myriad of intriguing and precious objects inside: magnifying glasses and spectacles clouded with time. Humble, leather-bound journals. Weathered tools and pottery, each piece telling a unique and fascinating story. I remember the feeling of being immersed in completely different worlds, learning about nineteenth-century farm life, Native American history on the plains, and beyond. These experiences served as a passport to history and ignited my imagination—to explore beyond what I thought was possible.

I built on my childhood curiosity by attending great colleges and eventually leading one, Colorado College. My exposure to different institutions of learning throughout the years—from museums and libraries to schools and colleges—changed my life and taught me the importance of using a blended approach, or a transdisciplinary framework, to better understand our interconnected world. Supporting historians, artists, economists, and biologists to come together with students and community members to respond to a local challenge is the magic of the liberal arts in action.

In *Interpreting Science*, editors Debra Reid, Karen-Beth Scholthof, and David Vail bring together expert insights, perspectives, and learnings to shed new light on the importance of integrated learning and the nexus between science and history. The book also explores how museums and historic sites can provide the perfect venue for transdisciplinary work and be catalysts for building awareness and inspiring change.

At the National Geographic Society, transdisciplinary approaches have been central to our mission for more than 135 years. This is evident in the depth and breadth of our work, from our earliest scientific expeditions and educational programs to our coverage in *National Geographic* magazine, our historic archival collections, and our Washington, D.C.–based museum and traveling exhibitions in partnership with local museums around the world. Over the decades, we've learned time and time again that when we weave together science, nature, education, history, and storytelling, we can engage and inspire more people, at a much deeper level—and importantly, galvanize interest and action.

Today, we drive scientific engagement and impact by funding a global community of National Geographic Explorers—leading scientists, conservationists, technologists, educators, and storytellers—who share our mission to illuminate and protect the wonder of the world. In pursuit of answers to the big questions, our Explorers work at the intersection of multiple disciplines by design—investigating, pushing boundaries, and solving challenges to advance the world's body of knowledge and make the greatest impact possible.

For example, our traveling interactive exhibition, *Becoming Jane: The Evolution of Dr. Jane Goodall*, celebrates the life and work of legendary ethologist Jane Goodall, who helped shape the history of the modern conservation movement while inspiring the next generation of scientists and environmental stewards. Another example is a group of geneticists associated with the National Museum of Egyptian Civilization who, supported by the Society, are extracting ancient DNA from Egyptian mummies to glean new insights that will help advance biomedical sciences. It's a story that's captivating the public's interest while teaching about science and history. A third example is geographer and Explorer Victoria Herrmann, who is bringing together local communities, the International Council on Monuments and Sites, and at least ten heritage sites globally to share resources, educate the public, and create plans to safeguard historical and culturally significant places, like Petra in Jordan, from the impacts of climate change.

As global issues like the climate crisis, public health threats like COVID-19, and social injustice accelerate, sharing transdisciplinary knowledge is more vital than ever to solve the complex challenges of our time. As leaders of local museums and historic sites, you and your institutions have an indispensable role to play. You serve as true cornerstones for your communities, informing, educating, inspiring, bringing people together and connecting them to the past, generation after generation. Through the close and trusted relationships that you've built with your constituents, you have a unique and crucial opportunity to catalyze social consciousness.

By incorporating more science into your exhibits, educational experiences and other storytelling materials, you can help build understanding of today's critical issues, like health and sustainability, while continuing your important mission work to illuminate the past. Adding science can also benefit your core programming, providing new ways to engage the public and tell your historical stories. By adding a scientific lens, museums and historic sites can have an even greater impact for local communities and for the planet—telling the stories of our past while inspiring people to do their part to create a more sustainable and equitable world.

In her thoughtful chapter in *Interpreting Science*, Melanie Armstrong says that experiences at museums and parks can "imbue science with the critical thinking and civic sensibility to transform information into understanding and identity." Now is the moment to tap into this power.

There's never been a more critical time to find creative, integrated ways to drive understanding and action and to build a global movement of changemakers. Museums and historic sites can be the example. Every time a visitor opens the door to one of your invaluable institutions, there's an opportunity to connect them with the past while opening their eyes to what they can do to help shape a better future.

Jill Tiefenthaler, PhD
Chief Executive Officer
National Geographic Society
March 10, 2023

Preface

Karen-Beth G. Scholthof

The artifacts, archives, buildings and landscapes, living collections, living memory, and cultural practices that museums collect and interpret offer unique opportunities to enhance science literacy. In fact, these historical resources bring the Two Cultures, history and science, together, and *Interpreting Science at Museums and Historic Sites* shows how to do this. Many may immediately associate this effort with STEM (science, technology, engineering, and math), STEAM (adding arts), and STEMIE/STEAMIE (adding innovation and entrepreneurship) programming. Convergence education, defined as the application of "knowledge and skills using a blended approach across multiple disciplines (i.e., transdisciplinary) to create and innovate new solutions," calls for this work. The history of science can ground this work, but *Interpreting Science* encourages readers to take the next step. Close study of collections in museums and historic sites helps us recognize the potential to connect history with science, and vice versa.[1]

Interpreting Science asks readers to think first about the untapped potential of historical resources to tell science stories. This is how we connect science to human experiences and vice versa. We can start by putting scientific milestones into a local history timeline. We then continue by documenting the ways in which historical resources increase science understanding, be it the routines of individuals who collected meteorological data, studied nature, or battled tuberculosis and lobbied for pure food and drugs. The coeditors and contributors explore topics that show how this works. All work across disciplines, and each has been informed by their engagement with museums, archives, and historic sites. They understand, firsthand, how science informs history interpretation and how this expands audience engagement and can prompt and sustain community conversations. Contributors avoid jargon, so readers who lack formal education in any given scientific discipline can still engage with the contents. Collectively, the chapters emphasize the urgency of this work and provide inspiration to get started. Individuals working in science, history, and art museums and at zoos, nature centers, arboretums, and aquariums can all benefit.[2]

Discovering Science

William Whewell, who coined the term "scientist" in 1833, provides us with a positive framework to interpret science at museums and sites. For Whewell, "discovery comprised three elements: the happy thought, the articulation and development of that thought, and the testing or verification of it."[3] This "happy thought" prompts a creative moment, inspires curiosity, and foments an experiment. This leads to analysis in the lab, field, or workshop, and perhaps, the development of an application. Of course, happy thoughts do not emerge out of thin air. They are based on experience and a worldview. Experiences that engender happy thoughts—experiences that prompt curiosity—may include formal learning (coursework and teaching) and informal learning (observation and apprenticeships) or the need to solve a particular problem (food, health, transportation, shelter).

What we wish to convey in this volume is that almost any object can be used to interpret this process. We encourage you to begin this process by identifying objects in your collection that inspire curiosity. Every chapter in this book models this practice, explaining and showing how science is participatory, accessible, interesting, and nearly infinitely adaptable within the themes of your museum or historic site.

How can you bring science to your site and engage visitors? Most of us have had little or no experience with formal science beyond an introductory college course. Yet, this should not limit opportunities to link historical resources to science. In fact, interpreting science does not require the expertise of a practicing scientist because experts may also be limited in their expertise. For example, we can generalize that a physicist will have little knowledge of mushroom cultivation. Similarly, an entomologist will have little knowledge of number theory. We stewards of historical resources know our collections, our museums and historic sites, and our communities best. This knowledge can translate into compelling, exciting, and relevant science interpretation. We also can gain confidence from *Interpreting Science* to reach out to science experts who can inform and enhance science stories that historical resources can tell.

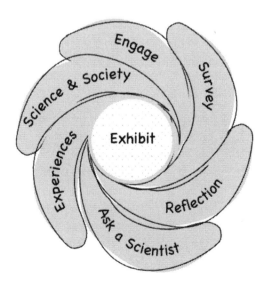

Figure 0.1. Science+History= Engagement. The petals on this flower illustrate distinct but interconnected steps to take to ensure that an exhibition brings science and history together.
Created by Karen-Beth G. Scholthof. Used with permission.

The petals on the flower illustrated in "Science+History=Engagement" serve as a checklist worth reviewing when planning exhibits, selecting artifacts, and writing labels. "Engage" should prompt us to establish a cogent, scientifically authentic orientation at the guest entry point to the Two Cultures. "Science and Society" compels us to identify a historical resource around which we build scientific and historical relevance. "Survey" reminds us to document guest perceptions via straight-forward questionnaires (accessible via a QR Code) at multiple stages in the process, at the beginning (formative evaluation), middle, and end (summative evaluation). Do visitors feel welcome to explore science and history? Does a particular object pique their curiosity? What can make guests more comfortable about asking science (or history!) questions? Survey findings can prompt ongoing proactive changes relative to artifact selection, label writing, or exhibition formation and inform future Science+History=Engagement.

"Experiences" ground engagement. Experiences ensure that guests are active participants where the Two Cultures come together, a place that is at the heart of the lived experiences of individuals and communities. "Ask a Scientist" (or "Ask a Historian") draws in content experts who can facilitate programming. This might take the form of a Science Café, Master Gardener and Citizen Scientist events, chemistry demonstrations, and public lectures, among others. The goal is to reinforce that science is a human experience and its outcomes are embedded within our social compact. "Reflection" offers a space for the guests to think about the role of science in society—and to consider how a particular science has affected an individual, family, or community. And beyond the local: consider the role of the citizen in voicing opinions about science, and the realization that science is rarely the black and white dichotomy of good or bad, beneficial or detrimental. Thus, the process should address not just what exists or what survived or what thrived. Museums and historic sites can address a particular theme or object by drawing on historic examples of doubt, resolution, or consequences of the cost to individuals and society when science was defunded, or advances and findings questioned. Reflection (collected as a public comment or survey response) can inform the curator to revise, build-on or re-interpret historical resources. Science+History=Engagement reminds us that this is an ongoing process of engagement, conversation, experience, and interpretation that can help us realize the full value of adding science to history at museums and historic sites.

Content and Context in Science

The framework for this volume was inspired, in part, by Peter Galison's "Ten Problems in History and Philosophy of Science." One problem, "What Is Context," brings us to interpreting science at museums and historic sites: "A focus on the practice of science—the structure of scientific work in the field, museum, observatory, classroom, or shop floor—offers the opportunity to address scientific concepts and comportment in specific sites and circumstances *without* hewing to a vision of what is 'truly scientific' or 'merely exterior.'"[4]

For our purposes, context tends more toward the tactile than the textual. Material in collections can be used to show how science is woven into everyday life. For example, contributor Sam White shares the inspiration he took from a slice of an ancient tree on exhibit at Klimahuset (The Climate House, Museum of Natural History, Oslo, Norway). This serves as a meaningful introduction to scientific and social topics. The tree rings are the timeline, with a few pins placed at "important" historical dates. As a visitor, this exhibit may push us to a moment of reflection and recognition that "time marches on," but it doesn't enhance or inform our knowledge about *this* tree, science, or history. Yet this tree can be used to facilitate a deeper, more interesting narrative of science and human experiences, such as interpreting changes in the climate over eons. For example, in 2023, Earth is home to eight billion humans and three trillion trees. How is it that humans have had such a profound impact on the Earth? What can a tree tell us about forests and ecosystems; the classification of species by their features and DNA; wildfires and climate; land management by Indigenous peoples; human, animal, and plant health; shipbuilding, home construction, and paper manufacturing; and environmental movements?[5] It turns out that the answer is quite a lot. The history of the Earth is embedded in *this* tree—both conceptually and literally. *Interpreting Science* encourages us to develop our own ideas of how to interpret science using everyday objects, and it provides the case studies of how to do that.

Ethics and Morals: Shifting Meaning in Science

A second pitfall, using Galison's terminology, in interpreting science is the "purity" of science. In part, this touches on the ethics and morality of science. Who makes these decisions? Are some outcomes impure, perhaps atomic weapons or genetically engineered foods? Does your location (regional, national), socioeconomic status, ethnicity, gender, and education affect your interpretation of what is pure or impure science? Another aspect of "purity" for scientists is the philosophical divide between basic (mostly laboratory) and applied (mostly field) sciences. Is laboratory research that is initiated without any intended application more "pure" than the research associated with developing a new dahlia variety? Is laboratory and field research at a university more "pure" (or trusted) than the same experiments and technology developed at a government or industry laboratory? This also extends into thinking about the sciences: Is physics a purer science than botany? Than psychology or math? And, how have our ideas of science changed in a decade or a century? Interestingly, science itself does not provide the questions or answers here. Rather, "purity" is a sociological and historical interpretation of science, technology, and its workers—the scientists. Artists also have key interpretations of science, for example, knitting mathematical shapes; and theater, music, and writing to interpret disease.[6]

Disciplinary Boundaries: What's in a Name?

Another aspect to consider is the disciplinary boundary, with discovery and technology being attributed to a particular science at a particular time. This also extends to individuals or groups of scientists who are keen to categorize themselves for prestige, forging new disciplines, and the "bandwagon" effect. This also intersects with the "purity" aspect of science in that discipline can have a certain cachet. For example, synthetic biology is hot, botany is not.

Gene editing of organisms is hot, collecting and classifying organisms is not. All, however, are aspects of biology. Some of this re-classification of scientific practice is due to increasing specialization, increasing costs of research (employees, materials, travel, publication) that require a focus on a particular field, and some is to gain the attention of funding agencies (or a shifting focus by the same agencies). What imbalance derives from inadequate research funding? How can society more evenly distribute scientific gains and mitigate the costs?

As knowledge accumulates, some areas of study become possible, others fall out of favor (no longer considered "cutting edge" or shunned due to real or perceived moral or ethical boundaries). For example, eugenics is no longer a field of study, and no biologist would define themselves as a eugenicist. How do categories define science and scientists in a particular period, and thus convey additional evidence of culture, ethics, economics, or politics? The work undertaken by a scientist also changes. Is a biologist working in the 1880s doing the same work as a biochemist (1940s), a molecular biologist (1980s), a bioinformatician (2000s), or a synthetic biologist (2020s)? Why are these titles changing over time? When interpreting science, the fields (or specializations) serve as a focal point to interpret why and how a particular science was known, seen, or popularized by the public.

The first step in contemplating these questions within museum research and interpretation involves locating an object that seems interesting. What was it intended for? Where was it developed? How did it travel as an object? How was it used? When did it become available, and who were the first recipients of the new science or technology, when did it fall out of favor, and why? [7]

Historical Conditions and the Emergence of Scientific Objects

An important consideration for understanding science is to determine under "which circumstances" might "collecting and observing together bring new kinds of objects or classes of objects into existence?" Galison gives several examples, to divine whether an object can be defined as "natural kinds" that, due to human agency, may no longer be "natural"—thus, a designation of "artificial kinds." Can an object derived from "nature" become "artificial"?[8]

Here, an example is uranium. The meaning of this element today is attached to mining and its downstream outcomes of environmental contamination, human health, and social justice and the meaning of Native American sovereignty.[9] During World War II and through the Cold War, the meaning of uranium was the Manhattan Project, atomic weapons, and nuclear winter. Yet, in 1879, a brief note in *Scientific American* celebrated the discovery of uranium in California. Now the United States would have its own source of this sought-after and valuable chemical element used to color glass and tint photographs.[10] Was uranium "natural" when it was in the ground, when it was studied as a chemical element? Did it become "artificial" when it was manufactured as an atomic weapon? Used for nuclear power? Is the uranium "natural" or "artificial" when it is declared an EPA Superfund site? In short, "what is the difference between the made and the found," or has the construct of "natural" and "artificial" always been artificial?

In this instance, a pretty green-glass uranium household object, a milk pitcher/creamer, reveals itself as a complex scientific story.[11] Uranium has a deeper meaning when viewed as an object for fundamental science experimentation in the areas of chemistry, then physics, then biology, and therapeutic medicine, public health, and environmental studies. The links for history and science induce the evaluation of the "pure" science of radioactivity and radiation compared with the "purity" of science in regard to nuclear weapons, nuclear energy, scientists associated with test-ban treaties, and the Bulletin of Atomic Scientists' "Doomsday Clock."[12]

Figure 0.2.   Block Optic green glass creamer by Anchor Hocking. Photograph by and collection of Sonia C. Irigoyen, PhD.

These examples are intended to be a first step toward finding and defining an object or site that is interesting to you and that has potential to enrich and enhance your museum through the lens of science. An exhibit may look at conversations related to science, how an object made its way to a historic site, the extant (natural) history of science at a site (plants, animals, soil, weather), and how people(s) were changed, awed, or influenced by science and technology. And, of course, there is room for interpretation of the science of today to compare and contrast our lived experiences with those on your site.

An Overview of *Interpreting Science*

This collection of chapters should help those with little formal science training identify and develop science interpretation in museums and historic sites. Each chapter encourages engagement with historical resources to deepen content knowledge. The intent is not to create scientist impersonators, but to launch and sustain conversations with scientists, special-interest groups, and the general public. The chapters do not address all sciences, but this results from lack of space rather than lack of potential.

Part I, Science, the Public Historian, and Museum Collections, addresses the responsibility that science and history museums share for the long-term stewardship of artifacts, archives, and science evidence. Contributors affirm that drawing science content out of historical resources can inform current issues and inspire audience engagement. Chapters on documenting climate history, assessing weather instruments, exploring medical science, and gaining insights through experimental archaeology indicate the potential of this work.

Part II, Science and the Human Experience, focuses on physical and social sciences inseparable from the human experience. Each chapter uses a topic (chickens, dahlias, tuberculosis, inadequate health care, irrigation and mountain snowpack, and food policy) to learn more about ourselves and about science.

Part III, Science: A Culture of Doubt, A Culture of Questioning, builds on ideas raised by Naomi Oreskes and Erik Conway in *Merchants of Doubt* (2010). Contributors explore the incongruity between public trust in museums and public rejection of scientific facts, the cost that society pays when science research is defunded or findings questioned, and the ongoing interpretive tensions around race, science, health, and activism. Topics include how to engage with problematic notions around "science as progress," the historical complexities of genetic engineering as well as reflections on race, health, and the weaponization of science, specifically related to the atomic bomb.

Part IV, Science and History Museum Education, explores minds-on activities that make history and make science inclusive and accessible. One chapter addresses the ways that art-science integration (STEAM) supports creativity, curiosity, close observation, and questioning. These, in turn, support innovation. Others address methodologies such as school gardens, nature study, and storytelling that draw attention to science practice. Close reading of yearbooks published by the US Department of Agriculture affirms the potential of untapped resources to teach about twentieth-century domestic engagement with science. Curriculum materials convey the physics, science, and technology lessons inherent in an exhibition on the history of automobile racing.

The concluding chapter urges us to make science and the humanities more inclusive. Suggested readings and a timeline provide resources to help you fit your museum or historic site into the long history of discovery and science. Use these to start an assessment of your institution's historical resources that inform our understanding of scientific topics.

In conclusion, coeditors and contributors ask you to ponder not what science can do for you, but what humanities can do for science. *Interpreting Science* emphasizes the urgency to bridge disciplinary disconnects. Humanities disciplines provide evidence that we need as we continue to wrestle with life in the approximate. We want answers, and many expect science to deliver those answers. How long will I live? What risks come with inventions? Will we be prepared for the next pandemic? Will we solve the climate crisis? Do we need to? Will we learn to coexist with the planet?

Science can provide some answers, but it is often easier to see these in hindsight. At present, it is difficult to face the uncertainty caused by the time lag that occurs between discovery and confirmation of societal benefit. Humanities can reinvigorate our connection to science, past and present. It remains a given that science will inform our future though how remains unknown. Future generations will have to assess the ramifications. Let's get started with this important work!

Notes

1. Several organizations drew attention to science and related subjects during the 1990s. This included revisions of National Science Education Standards and efforts to leverage for science, mathematics, engineering, and technology education (SMET; now STEM) by the National Council of Teachers of Mathematics and the National Science Foundation (NSF). Judith Ramaley, then director of the NSF's education and human resources division, coined STEM in 2001. See Jerome Christenson, "Ramaley Coined STEM Term Now Used Nation-wide," *Winona Daily News* (November 13, 2011). Comparable momentum did not materialize for humanities, though STEM's expansion to STEAM incorporated arts education. STEM's momentum caught the attention of staff in state and local history museums. Earth sciences and social sciences remain essential in National Curriculum Standards for Social Studies including civics, economics, geography, and history education, per the National Council for Social Studies. The 2018 Federal STEM Education Strategic Plan called out convergence education as one of four pathways to engage students in "real-world problems and challenges that require initiative and creativity." See *Convergence Education: A Guide to Transdisciplinary STEM Learning and Teaching: A Report by the Interagency Working Group on Convergence Federal Coordination in STEM Education Subcommittee*, Committee on STEM Education of the National Science and Technology Council (November 2022), quote p. 7.
2. Santiago Ramón y Cajal's *Advice to a Young Investigator* (1916; Cambridge, MA: MIT Press, 1999) revisited in a January 6, 2020, blog, "The Most Important Scientific Problems Have Yet to Be Solved," https://thereader.mitpress.mit.edu/the-most-important-scientific-problems -have-yet-to-be-solved. Ramón y Cajal emphasized the lag that occurs between scientific discovery and recognition of utility. Coeditors Reid, Scholthof, and Vail each believe that putting scientific "discoveries" into a timeline can help the public understand this time lag. *Interpreting Science at Museums and Historic Sites* confirms that the significance of scientific findings takes time to confirm, and that significance was not apparent then, nor will it be apparent now.

    History museums engage in science interpretation, i.e., Chloe Dye Sherpe, "Using Science, History, and Art to Interpret Climate Change," *History News* (Spring 2020): 14–19, describes how the Skagit Climate Science Consortium, Skagit Land Trust, Skagit Watershed Council, and Swinomish Indian Tribal Community all participated with the Museum of Northwest Art in La Conner, Washington, in exhibition development and delivery and conversations targeting shared concerns about climate change. Others address the need for science museums to engage with communities by touching a nerve, as Michael John Gorman explained in *Idea Colliders: The Future of Science Museums* (Cambridge, MA: MIT Press, 2020): chapter 1, p. 7; and Pablo Martinez, "Current Science in Museums and Science Centers" (master's thesis, University of Washington, 2016), confirmed that a lack of scientific knowledge made the general public hesitant to apply science-related issues to their own lives. His interviews showed that the museums followed some common practices to increase science literacy. Namely, they provided opportunities for the public to engage with scientists and staff and they rapidly changed exhibits to draw more attention to current science topics, including but not limited to climate change, medical research, or new technologies, all designed to make the public more confident in engaging with science issues directly.

3. This description in the *Stanford Encyclopedia of Philosophy* may take some liberty with Whewell's use of "happy;" yet, for our purposes, a "happy thought" is a low-stakes point of entry to interpreting historical resources within the framework of science. For more on the philosophy of scientific discovery, including the quotation, see Jutta Schickore, "Scientific Discovery," in *The Stanford Encyclopedia of Philosophy*, eds., Edward N. Zalta and Uri Nodelman (Winter 2022 Edition).

4. Peter Galison, "Ten Problems in History and Philosophy of Science," *Isis* 99 (2008): 111–24, quote 112, https://doi.org/10.1086/587536. Galison pointed out ten challenges that historians of science face, including discerning the meaning of context and the limitations of historical research to explain context. He also addressed ethical and epistemological distinctions between scientific purity (and impurity), the freedom to experiment (and moral and ethical limitations on experimentation), and natural things (compared to quasi-natural and artificial things).

5. On November 15, 2022, the United Nations Population Division reported the human population at 8 billion; https://www.un.org/development/desa/pd/events/day-eight-billion. In 2015, the global tree population was estimated at 3.04 trillion; see Thomas W. Crowther et al., "Mapping Tree Density at a Global Scale," *Nature* 525 (2015): 201–205, https://doi.org/10.1038/nature14967. A good introduction to human history is Yuval Noah Harari's *Sapiens: A Brief History of Humankind* (New York: HarperCollins, 2015). For more on the rich habitats in the forest canopy, see Richard Preston's *The Wild Trees: A Story of Passion and Daring* (New York: Random House, 2007).

6. For more on knitting, see Sarah-Marie Belcastro, "Adventures in Mathematical Knitting," *American Scientist* 101 (March–April 2013): 124–33, https://doi.org/10.1511/2013.101.124. Several examples of the arts and science themes include tuberculosis in Verdi's *La traviata* and Puccini's *La bohème*; HIV/AIDS in Jonathan Larson's rock musical *Rent* (itself a takeoff of *La bohème*); Tony Kushner's play and film *Angels in America*; and Alison Hawthorne Demings's *Science and Other Poems* (Baton Rouge: Louisiana State University Press, 1994).

7. *Scientific American* is a particularly good resource to follow science and technology in the United States from 1845 to the present. Scientific societies illuminate the boundaries of practice and members' evolving interests over time. Some questions related to professional societies include: When were they founded and why? For an example, see Paul D. Peterson and Karen-Beth G. Scholthof, "The Society That Almost Wasn't: Issues of Professional Identity and the Creation of the American Phytopathological Society in 1908," *Phytopathology* 100 (2010): 14–20. Were there international or national meetings? Were regional divisions of societies established, and when did they meet and why? Where were these societies headquartered? Who were the members (scientists in industry, the university, government)? Did the society publish member newsletters and journals? Do they have an archive open to researchers?

8. This section is heavily derived from Galison's "Ten Problems," especially pages 116–18. An example is vaccines. What is "natural" or "artificial": A potentially lethal childhood measles virus infection or prophylactic childhood vaccinations?

9. See, for example, "Navajo Nation: Cleaning Up Abandoned Uranium Mines," US Environmental Protection Agency, last modified July 18, 2022, https://www.epa.gov/navajo

-nation-uranium-cleanup; and Laurel Morales, "For the Navajo Nation, Uranium Mining's Deadly Legacy Lingers," *NPR Weekend Edition Sunday*, April 10, 2016, https://www.npr.org/sections/health-shots/2016/04/10/473547227/for-the-navajo-nation-uranium-minings-deadly-legacy-lingers.

10. See, "Uranium in America," *Scientific American* 41, no. 26 (December 27, 1879): 413, https://archive.org/details/scientific-american-1879-12-27/page/n5/mode/2up. In 1879, uranium ore was selling for $1,000 per ton—equivalent to $29,500 today. Green glass is colored by the addition of uranium oxide.

11. Milk pitcher/creamer, made of uranium glass, from the collection of Sonia C. Irigoyen, and photographed by her, appears in color on the cover images of this volume.

12. For the current time of the Bulletin of the Atomic Scientists Doomsday Clock see: https://thebulletin.org/doomsday-clock/current-time/.

Acknowledgments

We three editors approached this subject from distinct perspectives, but our training and experiences coalesced into a book that we believe serves a purpose. Namely, *Interpreting Science at Museums and Historic Sites* stresses the ways that humanities inform the sciences and the sciences inform the humanities. Being aware of each disciplinary approach will create vibrant and compelling programming. *Interpreting Science* inspires history museum staff to pursue science that relates to their mission and that draws on the expertise of scientists sympathetic to their cause. It urges science museum staff to incorporate humanities into their programming by drawing on the expertise of historians and other humanities practitioners sympathetic to their cause. The combination results in science programming that captivates audiences and draws them deeper into humanities and sciences.

Each of us put their mark on this book. Deb, the historian and museum curator, has little formal science training but has contemplated the categorization of history as either a humanities or a social science discipline for her entire career. She looks at everything through a history lens, and she monitored the museum barometer for this volume. Karen, the formally trained scientist and practitioner-historian, enthusiastically advocates for science history and includes humanities in her science teaching. She brought her deep knowledge and mastery of collegial critique to the project, and she kept her eye on the science barometer. David, a historian as comfortable with the environment as he is with agriculture, delivered content and sustained conversations that crossed scholarly, intellectual, and geographical boundaries to get at the humanities-based stories of the scientific past.

We three leaped into this project with many ideas and much enthusiasm, fanned by many years of conversation. Many contributors shared in these journeys, sometimes knowing one, two, or all three of us. It made for a vibrant collection of chapters, crafted by experts who met short deadlines with good humor and thoughtful response. It benefited from close reading by Meg Goldberg and Herman Scholthof. We thank Sumner G. Hunnewell for indexing this book. May this collection inspire you to think about science in your museum or historic site.

Debra A. Reid
Dearborn, Michigan

Karen-Beth G. Scholthof
College Station, Texas

David D. Vail
Kearney, Nebraska

March 14, 2023

SCIENCE, THE PUBLIC HISTORIAN, AND MUSEUM COLLECTIONS

The US population in mid-2023 is 350 million—every single person has depended and will depend upon science and scientists for basic needs and frivolous wants throughout their lifetimes. Here, the contributed chapters reflect on the role and responsibility that science and history museums *share* for the public understanding, meaning, and meaningfulness of science. Importantly, historical resources stand as concrete evidence of the development and use of science in a local setting. These collections invite the visitor to engage with materials and issues that are important to their communities. Chapters in Part I reveal that social and humanistic experiences—weather, disease, landscapes (homes, farms, archaeological sites), health and medicine—are enriched by historical queries as to how science has influenced our everyday lives.

While reading this section, note that each contributor describes a particular place at their site that is designed for the visitor to *reflect* on the meaning of the past and their current experiences. Time is an overarching theme in Part I. This includes large landscapes such as those described by Claus Kropp, whose investigations of early medieval (800 CE) farming practices at a UNESCO world heritage site reveals the science and meaning behind cultivation strategies. We, humankind, have been "doing" science for eons, a strategy not only to survive, but to find meaning and joy in our lives. Sam White, using a slice of a tree, and Melanie Armstrong, using a star party, explain how tree rings and telescopes can be used to simultaneously place ourselves in the past and present. They bring forth questions as to the meaning of time and how we can use objects to measure the years and light-years. Dark sky events are particularly meaningful because of the social interactions that we crave and use to place ourselves in the world. Roger D. Turner shows us how the seemingly mundane practice of decades-long daily weather observations can become a scientific record of climate change and help us understand the importance of collecting and interpreting data for scientific advances. April White and David Vail focus on public health in the mid-twentieth century, anchored in a tuberculosis hospital. They show us how to explore a disease and its effect on a local community—specifically to *feel* what it is to be human in suffering and remembering. White and Vail push us to extend our knowledge boundaries and to recognize that ethics and morality influence and can become the scientists' burden. Here, one role of the humanist may be to provide space at the museum or historic site to dissect a particular slice of science within its temporal period and to allow the visitor to reflect on how a particular science may have influenced their perceptions today.

CHAPTER 1

Communicating Climate Change with Archives of Nature and Archives of Societies

Sam White

THE FIRST display upon entering the Klimahuset (Climate House) set in the botanical garden of the Oslo Museum of Natural History is a cross section of an ancient tree trunk. Roughly two meters wide, it reveals annual growth rings from the tree's earliest summers in the 700s CE until its recent felling. A clear band along the horizontal axis marks the passing of each century, with pins marking the rings of notable years:

1349: The Black Death reaches Norway

1492: Columbus reaches America

1610: Lowest level of CO_2 in the Common Era

1740: Maximum of the Little Ice Age

1800: The Industrial Revolution

Similar displays attract viewers at museums throughout America and Europe, including a 1,500-year-old giant sequoia trunk in the American Museum of Natural History. In most cases, they serve as monuments to primeval forests of a bygone pre-settler or preindustrial era. Depending on the viewer's perspective and politics, they invite either regret for the destruction of natural splendor or celebration of human progress in the face of natural obstacles.[1]

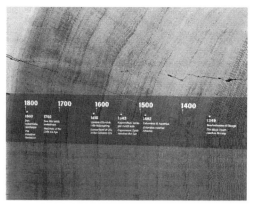

Figure 1.1. The first display at the Klimahuset, Natural History Museum, University of
Oslo, Norway. Photo by author.

Figure 1.2. Detail of timeline. Photo by
author.

The display in the Klimahuset serves a different role. The cross section of the tree visualizes both the passage of time and the variations of seasons, with wider or narrower rings
reflecting local temperature and growing conditions. The timeline not only represents the
events of human history alongside the life of the tree but also invites viewers to consider
both humans and trees as agents in a changing climate system, emitting and absorbing carbon dioxide. In this way, the display leads visitors into the first hall of the Klimahuset, which
explains the history and processes of climatic variability before the rapid global warming of
the present generation. It accomplishes all of this without lengthy text, charts, or graphs, or
heavy-handed messaging.

The chapter considers how and why the tree display at the Klimahuset communicates
climate change effectively. Recent studies have emphasized the challenges that climate
change poses for public science and especially science museums. Organizations and scholars
have recognized the roles that history might play in climate change communication. Never-
theless, science museums and public science could make more use of a central element in the
study of past climate and society: what this chapter will explain as an integration of "archives
of nature" with "archives of societies." This integrated presentation of human and natural
evidence has epistemic value as well as visual appeal, which aids in the communication of
climate change and its human impacts.

Climate, Public Science, and History

Recent studies point to climate change as both a potential crisis for public science and an
opportunity for institutions, including science museums, to reimagine their roles. As these
studies highlight, museum installations and similar projects are tasked with communicating
complex messages about climate science that can provoke political and emotional resistance:
for example, that climate change is a global problem, driven by greenhouse gas emissions,
yet its impacts are varied and local; or that the science on man-made warming is certain,
yet its speed and effects remain hard to predict. They are also faced with communicating

the magnitude of the problem without provoking panic or despair, as well as the urgency of political action, while keeping the science above partisan politics. Above all, these studies emphasize the need to *communicate*—that is, generate thought and engagement—rather than merely put out information.[2]

Public history and historical scholarship play acknowledged roles in this task. First, studies have examined how public communication of climate change can draw on local cultural heritage. For instance, cultural heritage landscapes can highlight baseline climate and environmental conditions, illustrate local change, and highlight climate change threats to historic locations. Local knowledge and traditions may preserve memories of extreme events and effective responses and may help people recognize and come to terms with a changing climate.[3]

Second, public history can provide stories of change and human experiences that render climate change more accessible, compelling, and meaningful. We are fast moving into a "no analogue" situation in terms of climate—that is, a world of high temperatures and extreme weather beyond the range of natural variability in centuries or millennia of recorded history. Yet as David Glassberg argues, there are still historical analogues for aspects of our present situation, including a long history of human impacts on the environment and environmental factors in history. There are parables of catastrophe and collapse, hopeful tales of resilience, and many historical examples that fall somewhere in between. Thus, public history can "build resilience by telling stories about the past that avoid determinism and restore a sense of contingency, intention, and . . . radical possibility to the present."[4]

Climate History: Integrating Archives of Nature and Society

Largely absent from these discussions, however, has been the field of climate history itself. Climate history is an interdisciplinary specialization that examines past climate variability and change, extreme weather, and their human dimensions, integrating methods from the humanities, social sciences, and natural sciences. The field has grown rapidly in the past two decades, spurred by new work in climate science and rising concern over climatic change.[5]

A characteristic feature of climate history has been the integration of two distinct sources of evidence, which historian Christian Pfister has termed "archives of societies" and "archives of nature." To understand these terms, we can imagine all the ways that weather and climate are leaving traces in the world around us right now. Perhaps people are out buying winter coats or at home cranking up their air-conditioning; perhaps leaves are falling or flowers blooming; perhaps lakes are freezing or evaporating in the heat. Some of these effects will leave measurable traces generations or centuries from now, enabling future researchers to reconstruct aspects of present climate: perhaps records of coat receipts or utility bills, social media posts about the weather, traces of pollen in soil, or layers of sediments at the bottom of lakes containing proxies of temperature and precipitation. These accumulated traces, whether man-made or natural, will become "archives" of climate information.

Famous "archives of societies" include weather diaries and almanacs, historic flood markers, and records of harvest dates; while famous "archives of nature" include ice cores and tree rings. Normally we think of these two types of archives as belonging to two different domains of knowledge and research: the former to history and humanities and the latter to natural sciences. In climate history, however, the two forms of evidence are complementary. Archives of nature bring continuous, homogenous, objective information, and archives of societies bring detail and experience at a human scale. In this way, climate history generates more accurate and specific reconstructions of past weather and climate; demonstrates and quantifies their impacts; and tells richer stories of past climates and societies—whether stories of crisis and collapse or survival and resilience.[6]

The effectiveness of this combination of evidence illustrates a deeper principle of epistemology. Most historical knowledge is grounded in explanatory inference (or what philosophers sometimes call "abductive inference"). It is a probabilistic reasoning from effects to causes, or, more specifically, from the traces of the past to the events and conditions necessary to explain those traces. When historians read old chronicles and conclude that a famine took place some year centuries ago, they mean that famine provides the only or the best explanation for the descriptions in the chronicle. When dendroclimatologists measure old tree rings and conclude that there was a drought, they mean that drought provides the only or the best explanation for those narrow rings. When climate history specialists examine these two forms of evidence together, they come up with more specific, detailed sets of events and conditions that may accurately explain both at once.

This principle has guided recent methodological work within the international Past Global Changes Climate Reconstruction and Impacts from the Archives of Societies (CRIAS) working group of climate historians.[7] It is also integrated into the CRIAS "Weathered History" project: an online public exhibition on climate and society from ancient to modern times, integrating archives of societies and archives of nature. As viewers scroll down the exhibition, they zoom in on manuscripts and flood markers, ice cores and tree rings, weather-related art and artifacts, and even old wine bottles. The impact of the exhibition comes from visualizing in detail the manifold traces of past weather and weather-related activities in both natural and human records, inviting viewers to imagine for themselves the historical climatic, societal, and cultural conditions that must have left them.[8]

Tree Stories

The same epistemic principle also helps explain the effectiveness of the Klimahuset exhibit and its lessons for climate change communication. The cross section with its distinct growth rings evokes knowledge of climate variability and change by prompting viewers to imagine the seasons that must have produced each ring—both warm seasons and cold, according to the ring's width. Embedding dates and historic events within the rings embeds human history within its environmental context. Were the timeline merely placed *next to* the tree rather than *on* it, much of the effect would be lost.

Moreover, the effect would have been lost entirely had the Klimahuset merely overlaid a timeline of human events over a timeline of reconstructed global temperatures. The tree cross section is not only visually appealing, with its impressive size and the fascinating detail. More importantly, the viewer intuitively grasps that these growth rings are the traces of past seasons. The historical dates serve as a reminder of the human presence and experience throughout these seasons. To explain what they are seeing—to make sense of the patterns in rings and dates—the viewer must imagine the passing of years that produced them, their intricate variations of weather and climate and of human-environmental interactions. Timelines or graphics of climate change, however well contrived, evoke different knowledge because they may invite different explanations. Those who understand and trust climate science will explain the information as a product of scientific investigation. They might imagine the scientific labor and expertise behind them and find their beliefs and understanding reinforced by the display. Yet climate change deniers will interpret the information as the product of activist curators or politically biased messaging, merely reinforcing their doubt or disgust. Archives such as tree rings, formed by centuries of natural processes that leave visible traces, cannot be explained away so easily. Combined with archives of societies, they evoke a rich history of human-climate interactions that circumvents conspiratorial or contrarian explanations.

Of course, the growth rings on a single trunk can tell only a limited story. It is the story of an individual tree's interaction with its climate and environment rather than the dozens of tree samples needed for a reliable regional temperature or precipitation reconstruction. It is also a story that ends abruptly with the tree's felling, and not one projected into a future of global warming. Yet these limitations hint at another strength in this display: the power of empathy. Counting the rings of one tree helps viewers connect to climate change through the life of a single organism and relate it to their own stories, whose futures are not yet written.

Notes

1. The author thanks Anne Birkeland and Dominik Collet at the University of Oslo. Research for this article was supported by the Swiss National Science Foundation (grant IZSEZ0_207195/1) and Ohio State University. This is a product of the PAGES-CRIAS working group.

   Valerie Trouet, *Tree Story: The History of the World Written in Rings* (Baltimore, MD: Johns Hopkins University Press, 2020); Katrin Kleemann and Jeroen Oomen, eds., "Communicating the Climate: From Knowing Change to Changing Knowledge," *RCC Perspectives: Transformations in Environment and Society* 2019, no. 4, https://doi.org/10.5282/rcc/8822.

2. See, e.g., Fiona Cameron, Bob Hodge, and Juan Francisco Salazar, "Representing Climate Change in Museum Space and Places," *WIREs Climate Change* 4, no. 1 (2013): 9–21, https://doi.org/10.1002/wcc.200; Karen Knutson, "Rethinking Museum/Community Partnerships: Science and Natural History Museums and the Challenges of Communicating Climate Change," in *The Routledge Handbook of Museums, Media and Communication*, eds. Kirsten Drotner, Vince Dziekan, Ross Parry, Kim Christian Schrøder (London: Routledge, 2019), 101–14.

3. See e.g., David Harvey and James A. Perry, eds., *The Future of Heritage as Climates Change: Loss, Adaptation and Creativity*, Key Issues in Cultural Heritage (New York: Routledge, 2015); ICOMOS Climate Change and Cultural Heritage Working Group, *The Future of Our Pasts: Engaging Cultural Heritage in Climate Action* (Paris: ICOMOS, 2019).

4. David Glassberg, "Place, Memory, and Climate Change," *The Public Historian* 36, no. 3 (2014): 17–30, https://doi.org/10.1525/tph.2014.36.3.17. See also Jeffrey Stine, "Public History and the Environment," in *The Oxford Handbook of Public History*, eds. Paula Hamilton and James Gardner (New York: Oxford University Press, 2017), 190–201, https://doi.org/10.1093/oxfordhb/9780199766024.013.10.

5. Sam White, Qing Pei, Katrin Kleemann, Lukáš Dolák, Heli Huhtamaa, and Chantal Camenisch, "New Perspectives on Historical Climatology," *WIREs Climate Change* 14, no. 1 (2023): e808, https://doi.org/10.1002/wcc.808.

6. Stefan Brönnimann, Christian Pfister, and Sam White, "Archives of Nature and Archives of Societies," in *The Palgrave Handbook of Climate History*, ed. Sam White, Christian Pfister, and Franz Mauelshagen (London: Palgrave Macmillan UK, 2018), 27–36, https://doi.org/10.1057/978-1-137-43020-5_3.

7. For more information on CRIAS, see https://pastglobalchanges.org/science/wg/crias/intro.

8. For more on "Weathered History," see https://artsandculture.google.com/story/lgVhMeB lg6zDLg.

CHAPTER 2

Creating Public Space for Complex Conversations

Melanie Armstrong

IN MY final interpretive program as a national park ranger, I invited Canyonlands National Park visitors to peer through a telescope at the moons of Jupiter and deep sky objects like the twisted Whirlpool Galaxy and the wispy Orion Nebula. After a brief presentation, I trained my twelve-inch Dobsonian telescope on Saturn and ushered visitors to peer through the lens at the beaming planet. When I heard viewers gasp and exclaim, "I didn't know you could actually see the rings!" I understood a bit more of what it means to be an interpreter, and also a bit more of what it means to be human. As an interpreter, I created an opportunity for people to connect their grade school planetary science knowledge to an experience of awe and a new way of understanding their own lives on this planet. While every resident of Earth has the ability to look up toward the sky, the opportunity to see through streetlights and smog to the stars and planets beyond is increasingly rare. So-called "dark sky places" seek to preserve the experience of seeing the night sky, while tools ranging from telescopes to mobile apps can turn any site into a place for learning astronomy.[1] Dark sky programs build science knowledge; done well, they transform how people understand themselves as residents of the cosmos.

The night sky program provided a telescope and a guide to help visitors see, but, more importantly, it created a place for people to access ideas and perspectives that are too often obscured by the hums and whirs of modern society. The systems through which people attain knowledge are changing rapidly, most recently accelerated by online delivery mechanisms that took hold during the COVID-19 pandemic. Data are readily available through the internet, where learners have easy access to every piece of information I shared about stars and planets and light pollution. When they exclaimed aloud at the rings of Saturn,

Figure 2.1. Setting up a telescope before Arches' international dark sky park celebration, September 21, 2019, with Student Conservation Association intern Christine engaging with visitors. Arches National Park, Grand County, Utah. Photograph by Martin Tow, National Park Service.

however, they participated in a collective experience of science, which was an act of community building and citizenship. Such experiences at parks and museums imbue science with the critical thinking and civic sensibility to transform information into understanding and identity. During a time when traditional mechanisms of learning in brick-and-mortar schools seem increasingly precarious, such public places play an increasingly important role in generating transformative ideas about science, self, and society. In these spaces, skilled interpretation can help people peer through the social trappings that obscure their vision in daily life, creating the possibility of accessing the awe-inspired ideas that will be vital to addressing the most challenging social issues of our day.

A few times in my career, I led trainings for new National Park Service interpreters, during which I made a claim that was met with equal parts enthusiasm and disbelief: I promised that every park interpreter can change the world. I then laid out an argument, backed by science communication research, contending that changing the world happens when we create places where diverse groups can gather together and then prompt them to talk to each other in distinctive ways. People enter parks and museums motivated by intrinsic desires to learn, eager to see something that will transform their understanding of the world and perhaps even motivate personal change. Through exhibits, interpersonal engagements, and programs, skillful interpreters create a framework for social and individual change.[2] As critical sites for public interaction, such places are foundational to building a public sphere where social transformation can emerge from engagement with science, history, art, music, and—importantly—fellow visitors in the space.

Socially transformative science interpretation must begin by rethinking the audience, particularly relinquishing assumptions about an ignorant public. Society does not need better science to solve wicked problems like climate change, but needs better ways of communicating. We also need to learn to rely on ourselves as individuals responsible for our own actions, as well as contexts that resonate with things we already value or understand.[3] Museums and historic sites are vital to providing rich social context that can build our cultural understanding, even as experiences in parks enable people to exercise self-reliance and understand consequences. In these interactive spaces, a learner has "the freedom to make decisions (incorrect as well as correct ones) and observe their results."[4] When building contexts that resonate, it is important to remember that in informal learning environments such as museums, people do not consciously set aside their social climate, as they might attempt to do in a more formal learning environment.[5] Rather, in the museum space, visitors simultaneously uphold both their attention to social context and to exhibit content.[6]

Such learning entwines knowledge with accountability, inviting audiences to attune to the learning experience as part of the work of identity making. In the early twenty-first century, the rise of dialogic approaches to interpretation attempted to enlist diverse audiences as cocreators in the stories of a site, sharing their lived experiences through reflection boards as well as interpersonal dialogue. This approach recognizes that every person who comes to a site to look at the night sky, for example, has a context for understanding universal values such as darkness, time, and beauty. Subsequently, when they participate in learning experiences collectively, visitors build a shared understanding of diverse social histories and cultures. This takes seriously the role of a site as a space to dialogue on issues of public importance.

Consider the case of climate change. Science has never before generated such quality data about a changing climate, and yet there is still social conflict about key issues surrounding environmental change. Matthew Nisbet and Dietram Scheufele (2009) point to a myth that sits at the root of conflict over science. We assume that if people had "all" the information, they would see the "right" answer. We assume that controversy will go away when people are "brought up to speed." According to this deficit model, where the public suffers because of a perceived deficit in information, science communicators can be blamed for conveying bad information or the public can be blamed for not understanding the facts. The deficit model also assumes facts will be interpreted by all audiences in similar ways, when in reality people continually learn the technical aspects of science while also experiencing an evolving social, ethical, and economic context surrounding science. Most impactfully, the deficit model narrowly defines the range of problems that can be addressed, identifying a need for "more" or "better" science, while restricting the ability to integrate climate science with its social context, including the lived experience and value systems of the public. Conflict emerges from attempts to shift the framing of climate change beyond a science deficit model into a topic for public discourse.

As agents of the public sphere, parks and museums are uniquely positioned to bolster communication processes, including science interpretation, that provide resonant contexts and instill social responsibility. Integrating climate change into interpretive experiences at historic sites generates a context for climate science to resonate with prominent social values entwined with our public histories. Visitors can engage with a controversial issue outside the deficit model, replacing time spent acquiring endless amounts of information with the opportunity to wrestle with the social context that shapes how they understand and act upon the information they receive. Such sites also afford a vital opportunity to engage such controversial topics outside elite relationships of power.

While we readily recognize museums and historic sites as serving the public, we perhaps fall short of absorbing the complexity and importance of such public spaces. Using the case of Europe in the eighteenth century, the philosopher Jürgen Habermas theorized that certain historical factors enabled the creation of a public sphere.[7] During this era, an emerging merchant class began to gather in lodges and coffeehouses, where they exchanged knowledge and ideas along with goods and money. Such engagements led people to recognize their collective and individual human rights and spurred them to act on behalf of their own well-being, even to the extent of revolutions. Building spaces where people of diverse politics could simply gather and talk changed the means of creating knowledge, in turn developing new understandings of justice and human rights.

Parks and museums have similar commitments to public service at their core. This comes both in creating public access and in public-serving missions. In Habermas's words, "We call events and occasions 'public' when they are open to all, in contrast to closed or exclusive affairs."[8] Beyond guarding open access to knowledge and information, such public spaces make democratic processes possible. Hannah Arendt beautifully argued that the public sphere is "the common world" that "gathers us together and yet prevents our falling over each other."[9] We see this daily at our sites. Museum conversations are unlike the dialogues heard in restaurants, schools, or corporations; like merchants of yore, exchanges between global visitors at a site bring diverse cultural contexts to bear on a common topic at hand. Amid the chaos of a busy museum on a holiday weekend, one can witness the gathering of citizens with wide-ranging histories who are learning not to fall over each other, whether through direct conversation or through the sensory experience of witnessing how others engage with the space and topic. Do we fully realize the potential of these sites as rare spaces for active public engagement? Habermas predicted a decline of the public sphere due to mass media and the co-option of public spaces by the state. As enduring elements of the public sphere, museums might revitalize spaces where diverse citizens converge to deliberate the issues that impact their shared futures.[10]

Museums and historic sites do not simply memorialize and preserve; as harbingers of the need for social reform, they play a role in creating the future in which we want to live. Similarly, science and technology engage in future-making. The public sphere, then, is where science becomes political action. Motivated by intrinsic personal desires to learn, the public step into these conversations simply by visiting such sites.[11] By presenting science boldly in spaces that help visitors access their social values, these sites can guide people's learning and promote a vital type of civic engagement, changing the world by creating informed citizens committed to justice and democracy.

Notes

1. The International Dark Sky Places program promotes stewardship through lighting policies and public education. Designations recognize parks and reserves, but also Urban Night Sky Places and communities. See www.darksky.org. Sites like NASA's Sky Explorer (https://solarsystem.nasa.gov/skywatching) provide sky-watching calendars as well as education activities.

2. The study by Stern et al. on interpretive programs noted that interpreters who identified an intention to influence behavior were more likely to influence their audience toward behavioral change. Marc J. Stern, Robert Powell, Emily Martin, and Kevin McLean, "Identifying Best Practices for Live Interpretive Programs in the United States Park Service," Project Report for the US National Park Service, August 6, 2012, https://media.clemson.edu/cbshs /prtm/research/interpret-environ-education/Stern-Powell-Interp-Study-Aug-6-2012.pdf.

3. Matthew C. Nisbet and Dietram A. Scheufele, "What's Next for Science Communication? Promising Directions and Lingering Distractions," *American Journal of Botany* 96 (2009): 1767–78; Matthew C. Nisbet and Chris Mooney, "Policy Forum: Framing Science," *Science* 316 (2007): 56.

4. Richard Burton and John Seely Brown, "An Investigation of Computer Coaching for Informal Learning Activities," *International Journal of Man–Machine Studies* 11 (1979): 5–24.

5. Paulette M. McManus, "It's the Company You Keep . . . the Social Determination of Learning-related Behaviour in a Science Museum," *The International Journal of Museum Management and Curatorship* 6 (1987): 263–70.

6. Sherman B. Rosenfeld, *Informal Learning in Zoos: Naturalistic Studies of Family Groups* (Berkeley, CA: University of California Press, 1980).

7. Jürgen Habermas, *The Structural Transformation of the Public Sphere,* trans. Thomas Burger and Frederick Lawrence (1962: repr., Cambridge, MA: MIT Press, 1989).

8. Habermas, *Structural Transformation*, 1.

9. Hannah Arendt, *The Human Condition* (Chicago: University of Chicago Press, 1958), 52.

10. Notably, while museums endure, the experience of budget cuts and diminishing resources calls us to be ever more mindful of how economic and social privilege gives access to those who want to participate in this part of the public sphere. Fees create barriers to these unique public spaces, which are enriched by the inclusion of visitors from wide-ranging social contexts.

11. Aleck C. H. Lin, Walter D. Fernandez, and Shirley Gregor, "Understanding Web Enjoyment Experiences and Informal Learning: A Study in a Museum Context," *Decision Support Systems* 53 (2012): 846–58, https://doi.org/10.1016/j.dss.2012.05.020.

CHAPTER 3

(Re)constructing the Past

Research and Science Interpretation in Experimental Archaeological Open-Air Museums

Claus Kropp

EXPERIMENTAL ARCHAEOLOGY is a relatively new branch of applied research that studies objects, puts them in use, and then vividly communicates the knowledge gained from that scientific research to a broader public. Open-air museums and historic sites often apply the techniques of experimental archaeology.[1]

The core aspect of experimental archaeology is the scientific method that tests hypotheses inspired by archaeological findings. Oftentimes archaeological evidence is too fragmentary to understand the unearthed objects, or to support an accurate (re)construction of an object, or work routines or cultural practice. Sometimes the past is utterly "dead and gone,"[2] and we can try to approach it only with scientific approaches. Curiosity drives experimental archaeology, leading to hypotheses around which sites and museums design their research plan. Findings unearthed during an excavation of an Iron Age storage pit might prompt researchers to ask: What construction details help us understand how Iron Age dwellers protected their food supply?[3]

The methodological orientation of an archaeological experiment applies the same empirical standards as those of the natural sciences. Thus, we subject our archaeological hypotheses to practical tests. Repeatability and several investigation parameters factor into the research plan.[4] If the subject of experimental investigation is a (re)constructed Iron Age storage pit, for example, this would require installation of sensors at different levels of the pit (determination of variations in the microclimate) and documentation and analyses of the soil and the content inside the storage unit. The more that is known about the framework and environment relevant to the subject of investigation, the more precisely a hypothesis about the efficacy of construction methods can be tested.

According to Peter Reynolds,[5] experimental archaeology applies to at least five subject areas. Each prompts specific hypotheses and distinct research plans, and site staff and audiences engage with each in different ways (key to the public dimension of experimental archaeology). First: Structures (i.e., archaeological features such as houses, storage pits, etc.) stand out, figuratively speaking, as critical to (re)constructing a historic site. Second: Processes and Functions (e.g., the functionality of a symmetrical plowshare or the exploration of patterns of wear evident on a plow-knife [coulter]), confirm life's routines, essentially ephemeral except for the intervention of material evidence. Third: Simulations (e.g., knowledge gained in a long-term experiment about house decay) confirm evidence of transformations that ultimately result in an archaeological feature. Fourth: Probability Trials (a combination of aforementioned approaches, simulations that take structures and processes into account) confirm the rationale for historic life routines. A probability trial might focus on cultivation techniques used for historic grains with the goal of documenting yields, analyzing yields over time, and confirming variations based on technology used. Fifth: Technological Innovations (this refers primarily to the experimental development of technical aids for archaeological field research), specifically, tools and equipment (re)created in keeping with research findings and then deployed in field research.

A sixth distinction of experimental archaeology relates to the public nature of the laboratory where research is conducted. The research process, in fact, forms the basis for public engagement. The public can observe the process by which experimental archaeology creates knowledge through applied research that is more relevant, more immediate, and more interesting to the public.[6] The following case study of the Lauresham Open-Air Laboratory for Experimental Archaeology makes clear the invaluable role that science methodology plays in research and interpretation.

The Open-Air Laboratory is the didactic part of the UNESCO World Heritage Site Lorsch Abbey (Germany). The laboratory is a scale one-to-one model of an early medieval manorial site focusing on the Carolingian era (ca. 800 AD). It is built to function as an educational site for visitors to get a vivid and lively impression of that era, a sense of how buildings might have looked, and an impression of how past agricultural systems might have functioned. In addition, the site gives the visitor a sense of the possible landscape of the past: agricultural fields, meadows, gardens, livestock, and plants. At the same time Lauresham is also a serious research facility dedicated to experimental archaeology.

An ongoing research project focused on early medieval "ridges and furrows" provides a baseline here. Historians and archaeologists have attributed the undulating pattern of ridges and furrows on rural landscapes to a specific historic plowing technique. But have these patterns on the landscape a deeper meaning? Experimental archaeology indicates that the slight differences in elevation of ridges (high spots) and furrows (low spots) may have helped ensure more consistent harvests adequate to feed farm families, landlords, and ecclesiastical orders.

Researchers formulated the ridges and furrows experiment by accounting for all available sources documenting the past agricultural system. They collected written evidence of field conditions, technological innovations such as plows (e.g., the heavy moldboard plow), and landscape and field surveys. This included archaeological and iconographic evidence and ethnographic comparative studies derived from landmark publications such as *Tools and Tillage*.[7] This data became the basis for clear research questions based on "how" and "why":

> How do growing conditions differ for crops in furrows and on ridges?
> How does the growing condition relate to crop yields?
> What yield increases can be achieved through organic fertilization?
> Are there nutrient enrichments/shifts within the ridge-and-furrows system over time?
> What is the relationship between effort (of creation and maintenance of this cropping system) and yield?

The research plan designed to address these hypotheses incorporated a series of field trials. As researchers implemented agricultural processes under controlled conditions, they acquired additional scientific data on the soil climate (e.g., soil humidity and soil temperature), furrow depth, weather patterns, soil nutrient levels (and changes over time), and yields linked to field locations. These parameters resulted in data of interest to both the scientists and the visitors.

Figure 3.1.   Draft measurements on (re)constructed ridges and furrows in the Lauresham Laboratory for Experimental Archaeology (Germany). Photograph courtesy of Staatliche Schlösser und Gärten Hessen.

In the practical phase of the experiment, the interplay of the historical framework and the integration of modern scientific parameters finally allowed for some interesting observations, which proved pertinent to today's sustainability discussions. Specifically, the interplay of the characteristic ridges and furrows on the arable land proved, according to the yield analysis and the growth monitoring of the grain, to be a clever risk-spreading mechanism deployed by medieval peasants.

The experiment proved that in particularly wet years, the grain yield on the ridges was twice that of the waterlogged furrows. In a drought period, on the other hand, the result was reversed and the yield in the furrows was twice as high as the cultivation on ridges. On average—and in combination with a broad crop rotation—this system sustained grain yields even in years of extreme weather. This resulted in a more reliable food supply. These results were confirmed with soil moisture sensors that continuously collected data sets from zero to thirty-five inches in the ground and did so for both the ridges and the furrows. During a drought period the plants on the ridges were already out of moisture at depths greater than fifteen inches. In contrast, in the furrows, the moisture level remained sufficient within ten inches of the surface. In combination with the knowledge of the actual root lengths of the respective crops, this resulted in a quite clear picture of how this field system would inter-correlate with the atmospheric conditions.

Simplified, this field system would therefore allow for an average harvest no matter the overall weather conditions. This scenario does not, however, account for the actual peasant workforce and their draft animals available for field preparation, maintenance, and harvest. That depended on the respective landlord and the size of the peasant households as well as the regional availability of natural resources.

To conclude, experimental archaeology offers valuable insights in the micro level of a medieval agricultural system. This is in contrast to what is typically derived from written sources that often focus only on the macro level, such as the organization of manors or the services of peasants. All of these experiments are visible features for visitors to explore (or observe) during their visit at the Lorsch Abbey. By being able to literally look over the shoulders of researchers, visitors see firsthand the relationship between hypotheses, applied research, and historical understanding. The Two Cultures (science and humanities) collide as staff test hypotheses and discern meaning from the evidence. They translate this into site interpretation that links structures, tools, and life's routines to human need (food, clothing, and shelter) and community survival.

Visitor studies conducted at Kloster Lorsch confirm that guest attention spans increased as they engaged with experimental archaeology. Probability trials helped them connect the dots between unanswered questions (hypotheses) and more nuanced and complete under-standing of the site's history. This resulted in longer site visits and more guest engagement with historic content, both indicators of successful site interpretation.

Further Reading and Digital Resources

Jodi Reeves Flores and Roeland Paardekooper, eds., *Experiments Past: Histories of Experimental Archaeology* (Leiden: Sidestone Press, 2014).

Linda Hurbombe and Penny Cunningham, eds., *The Life Cycle of Structures in Experimental Archaeology: An Object Biographical Approach* (Leiden: Sidestone Press, 2016).

Lorsch Abbey (Germany), available at https://exarc.net/members/venues/lorsch-de (accessed December 10, 2022).

"06—Projektvorstellung Landwirtschaft im Experiment [Project Presentation: Experimental Agriculture]," Freilichtlabor Lauresham YouTube Channel (May 9, 2020), available at https://www.youtube.com/watch?v=7UA8R9hLcXo&t=25s.

"Lauresham Digital—Ridge and Furrow Research Project," Freilichtlabor Lauresham YouTube Channel (March 19, 2020), available at https://www.youtube.com/watch?v=SOITMhfsYtc.

"Pit-House Project," Freilichtlabor Lauresham YouTube Channel (March 26, 2020), available at https://www.youtube.com/watch?v=wMghrpxLdUc.

Notes

1. Roeland Paardekooper, "Experimentelle Archäologie: Wer macht's und was bringt's?" in Claus Kropp, ed., *Laureshamensia. Forschungsberichte des Experimentalarchäologischen Freilichtlabors Karolingischer Herrenhof Lauresham*, Vol. 3 (2021), 6–17. John Coles, *Erlebte Steinzeit. Experimentelle Archäologie* (Munich: Bertelsmann Verlag, 1973).

2. Matthew Johnson, *Archaeological Theory: An Introduction*, second ed. (Chichester, UK: Wiley Blackwell, 2010), 1.

3. Barbara Scholkmann, Hauke Kenzler, and Rainer Schreg, eds., *Archäologie des Mittelalters und der Neuzeit. Grundwissen* (Darmstadt: WBG, 2016), 127.

4. Peter Reynolds, "Butser Ancient Farm, Hampshire, U.K.," in *The Constructed Past: Experimental Archaeology, Education, and the Public*, edited by Peter G. Stone and Philippe G. Planel (London: Routledge, 1999), 124–35.

5. Reynolds, 128–30.

6. Paardekooper, "Experimentelle Archäologie," 12–14.

7. Axel Steensberg, Alexander Fenton, and Grith Lerche, *Tools & Tillage: A Journal on the History of the Implements of Cultivation and Other Agricultural Processes* (Herning, Copenhagen: Gad, 1968–1995).

CHAPTER 4

Local Weather, Distant Connections

Interpreting Meteorological Instruments and Data

Roger D. Turner

IN THE lull between Thanksgiving and Christmas, 1951, the US Secretary of Commerce wrote to a local woman in Brownwood, a town of twenty thousand in the north Texas hill country. Secretary Charles Sawyer thanked Pearl Smith for her forty-six years of service as a voluntary weather observer. Smith had written down measurements made on the thermometer and rain gauge in her yard, twice every day for something around seventeen thousand days (presumably minus time for vacations and illness, for which Smith would have been responsible for recruiting a replacement). Each month she mailed a standardized form filled with her table of numbers to the government Weather Bureau. The observations she and other people made were "in constant use," Sawyer wrote, "helping to solve problems concerning the conduct of our industry and commerce, our agriculture, and our national defense."[1]

Sawyer's thank-you note suggests the opportunities available in interpreting weather and climate records. Weather is everywhere, pervasive yet intensely local. Climate too is local, though increasingly framed in discourses of global connections. Records of local weather are pretty common in historical collections, and they are often visually dull: tables of numbers occasionally enlivened by a stray remark on the day's events. Nor is it obvious how to tell compelling stories about the instruments used to measure weather; some are beautiful, treasured objects while many have been designed to be as indistinct from each

Figure 4.1.   Pearl Smith using a rain gauge in her yard in Brownwood, Texas, ca. 1951. The slatted stand at right houses a thermometer installed by the U.S. Weather Bureau. Photograph courtesy of Roger Turner.

other as possible. Yet the atmosphere has profound effects on how we eat and drink, breathe and build, travel and fight. Scientists use historical records to understand how climate is changing and to test their predictions about how the atmosphere behaves. Showing how diverse people have engaged with the atmosphere may be a useful strategy for helping the public face the climate crisis. This chapter lays out three strategies for interpreting local weather records.

Networks of Observers

One strategy for interpreting local weather records is to explore the connections between local weather observers and national, even global, networks. Many local weather records have been created as part of larger projects to understand nature, control territory, or support economic activity.

Side by side with the invention of the key meteorological instruments during the seventeenth century, natural philosophers were interested in using weather observations to understand and compare the natural history of places. Robert Boyle invited gentlemen to acquire barometers and thermometers to record the patterns of the air around them and to submit these observations to the Royal Society for comparison and publication. Robert Hooke published a handbook for weather observers, suggesting how to maintain instruments and what kinds of observations to record. Even as early as 1676, these exchanges of local observations were used to argue about climatic changes perceived in the context of colonial changes in land use in Ireland and North America.[2]

Such prominent American Enlightenment figures as Thomas Jefferson and James Madison kept weather diaries using their personal meteorological instruments. Jefferson bought a thermometer in Philadelphia while writing the Declaration of Independence, then a barometer a few days after the Declaration was signed. He kept up an almost unbroken diary of weather observations until 1816.[3] Jefferson also badgered his colleague James Madison to stick to the task. Madison's decade-long weather diary is preserved at the American Philosophical Society (APS) in Philadelphia, where it has been digitized and transcribed. The APS used excerpts from Madison's diary to introduce the history of American climate science in the exhibit *Becoming Weatherwise*, which opened in 2022.

Meteorological instruments could be expensive, and scientists have long realized that the value of quantitative weather records depends on an extensive and continuous run of measurements using instruments that are calibrated and standardized, so they can be easily compared. During the nineteenth century, these limitations were overcome by organizations interested in nationalism and state building.

The US Surgeon General began requiring army surgeons to record the weather in 1814, in hopes of being able to connect outbreaks of disease to features of weather or local climate. In a noteworthy early example of what is now called "citizen science" or more inclusively, "community science," the Smithsonian Institution organized a voluntary observing network. By 1849 the Institution had loaned standardized instruments to 150 observers spread across the United States. Beyond the meteorological data produced, this project enrolled a far-flung citizenry into a project of creating an understanding of "American" weather, as they took directions and telegraphed their observations to Washington, D.C., for the construction of national weather maps.[4]

Local weather and climate observations were joined to the expansion of governance capability seen in many nations during the latter half of the nineteenth century. Governments around the world established national weather services between the 1850s and the 1880s. The US Weather Bureau began in the Army Signal Corps in 1870, with uniformed observers using the telegraph to report observations on the activities of Indigenous groups and striking workers as well as storms and air pressure.[5] In 1890, Congress established the Cooperative Observer Program whereby volunteers used government-supplied instruments to supplement the weather observations made by Weather Bureau employees. By 1895, more than two thousand volunteer observers were reporting to Washington, including many from territories that had not yet become states. During the 2010s the National Oceanic and Atmospheric Administration circulated a traveling exhibit playfully titled *Treasures of NOAA's Ark* to interpret objects from this history.

These accumulated weather observations became important within projects of settler colonialism. Climate statistics were used to attract settlers and boost commerce, by illustrating the quality of local climates for agriculture, for instance. Boosters of Denver, Colorado, used measurements of the city's wind to tout its airflow, arguing for its healthful climate compared to the still air of tubercular East Coast cities. As historian Greg Mitman has put it, "regional economies of health were built upon climate and sunshine."[6] Los Angeles employed a meteorologist named Ford Carpenter in the 1910s and 1920s to produce an annually updated brochure, *The Land of the Beckoning Climate*, featuring tables of average temperatures and rainfall to attract new residents and new industries, including moviemaking and aviation.

By showing how local observations were connected to national networks, history museums can show how science done in local contexts contributed to large-scale transformations like the construction of nationalism and new political identities.

Recovering Science by Suppressed Groups

A second strategy for interpreting local weather observations is to closely examine who was making them. This strategy offers a way to recover science done by minoritized groups, contributing to widespread efforts to diversify STEM.

Weather diaries and daily climate records can offer a way to show historical women as knowledge makers. The weather diary of Margaret Mackenzie had been largely forgotten in the archives of the Royal Society of Edinburgh until it was brought to scholars' attention in the early 1990s. With weather records as well as observations on plants, animals, and the seasons covering 1780–1805, this journal offered useful information for climate reconstruction projects as well.[7] In the United States, the American Philosophical Society has worked with volunteers to transcribe a weather diary written by Ann Haines of the Germantown neighborhood of Philadelphia during the middle of the nineteenth century. The APS framed this as a community science project—with both Ann Haines as a community scientist and the volunteers transcribing her diary as contributors to a contemporary community science project to recover historical weather observations. Excerpts from Haines's diary were displayed as part of the *Becoming Weatherwise* exhibit. From this chapter's opening anecdote, Pearl Smith represents one example of thousands of women who participated in meteorology as cooperative observers, often in decades-long work. In eras when expectations and societal structures drastically limited women's participation in the public sphere, home-based observing offered women one way of contributing to the production of scientific knowledge, as well as strengthening national defense and the economy, as the Secretary of Commerce claimed.

A more famous cooperative observer is George Washington Carver, the African American chemist and botanist who worked at the Tuskegee Institute (Alabama). Carver served as a cooperative observer for the Weather Bureau from 1899 to 1932. The NOAA Central Library has preserved and digitized the hundreds of copies of Weather Bureau Form 1009 that Carver used to record his observations. While recording the weather was not central to Carver's career, the weather station at Tuskegee represented recognition and a sign of respect during a time when the federal government had retreated from supporting the civil rights of Black citizens. Indeed, in the midst of Carver's contributions as a cooperative observer, President Woodrow Wilson segregated the federal civil service, eliminating thousands of middle-class and professional jobs for Black people. NOAA preserves and digitizes these records as part of its effort to produce a climate database. The NOAA Central Library has created a Web page to highlight the observations made by Carver and later observers at

Tuskegee from 1899 to 1954.[8] While NOAA chose not to explore the dilemmas of doing science for the federal government as it implemented Jim Crow, history museums might choose differently.

The urge to record weather sometimes leads to stunning objects. Historian Fiona Clare Williamson has written about weather observations made by British meteorologists captured in Hong Kong during World War II. While imprisoned in POW camps, these men recorded observations on whatever scraps of paper they could obtain, including colorful wrappers and the backs of cards inserted in packs of cigarettes. Today the cards are preserved by the Public Records Office in Hong Kong.[9]

Using Historical Records to Understand Environmental Change

Another strand of Fiona Clare Williamson's work illustrates a third potential strategy for interpreting local weather and climate records. Weather records can be used to show connections between local and global environmental change. Local observations can help both scientists and museum visitors better understand the current impacts and potential consequences of climate destabilization.

Williamson has brought her expertise in the history of science to make historical records accessible to climate scientists through the Atmospheric Circulation Reconstructions over the Earth (ACRE) initiative.[10] This collaborative international initiative recovers historical measurements of weather and translates those observations into data that can be used in three-dimensional reconstructions of past weather called "reanalyses." Climate scientists use these reanalyses to check their computer models via a process called "hindcasting," as well as to interpolate past climate conditions for locations where direct observations were not made.

Historical geographers like Cary Mock mine historical weather records to understand hurricanes that occurred before formal networks for observing and recording them. Informal observations can capture information that can be used to estimate the path, size, and intensity of tropical storms. Data from ship's logs and newspaper reports as well as letters, diaries, and other unpublished sources have been incorporated into the National Hurricane Center's North Atlantic Hurricane Database (or HURDAT), another re-analysis project that has helped meteorologists assess changing patterns of risk created by tropical cyclones.

In Britain, the geographer and environmental historian Georgina Endfield uses historical documents and informal records to understand how people have adapted to extreme weather conditions in the past. From 2013 to 2016, she led a research project to create the TEMPEST database.[11] The project has mined archival documents including letters, diaries, church records, and school logbooks to build a database recording five hundred years of extreme weather in the United Kingdom.

Finally, ecologists have increasingly looked to historical records to understand how plants and animals are adjusting to a changing climate. Many weather diaries include phenological data, meaning observations of seasonal changes like dates when plant species start to bloom, when birds return from migration, or when hibernating mammals emerge from their dens. Perhaps the most famous of nature journals is Henry David Thoreau's observations of Walden Pond recorded between 1852 to 1858. Comparing flowering times then with flowering times today has confirmed that flowers bloomed in the eastern United States during the Springs of 2010 and 2012 at record-breakingly early times.[12] Since many organisms depend upon the timing of phenological events, such as birds arriving back at breeding grounds just as protein-filled insect populations swell, understanding changes in phenology is essential to our understanding of how much climate change species can adapt to without catastrophic losses.

Conclusion

Local weather observations may not be an obvious choice for an exhibit. They're often full of tables of numbers dense with information or hand-scrawled journals with little emotional content. But they offer local museums a way into stories on a grand scale. From the creation of national identity to stories of global exchange, weather and climate data can be surprisingly powerful. Local observing has offered minoritized people a way into knowledge production, which can connect historical figures to contemporary community science projects. And those historical records provide unique evidence from the past, offering today's scientists insight into how our actions destabilizing the climate today will diminish our future. Look through your collection and take a chance on displaying weather's history.

Notes

1. Charles Sawyer to Pearl Smith, November 27, 1951. Letter in author's personal collection.
2. Brant Vogel, "The Letter from Dublin: Climate Change, Colonialism, and the Royal Society in the Seventeenth Century," *Osiris* 26 (2011): 111–28.
3. Jay Lawrimore, "Thomas Jefferson and the Telegraph: Highlights of the U.S. Weather Observer Program," *Beyond the Data Blog*, NOAA Climate, May 17, 2018, available at https://www.climate.gov/news-features/blogs/beyond-data/thomas-jefferson-and -telegraph-highlights-us-weather-observer.
4. Eric Holthaus, "The National Weather Service Began as a Crowdsourcing Experiment," *Smithsonian Magazine*, October 2021, https://www.smithsonianmag.com/science-nature /history-national-weather-service-180978585/.
5. James R. Fleming, "Storms, Strikes and Surveillance: The US Army Signal Office, 1861–1891," *Historical Studies in the Physical and Biological Sciences* 29 (1999): 315–32.
6. Gregg Mitman, "Geographies of Hope: Mining the Frontiers of Health in Denver and beyond, 1870-1965," *Osiris* 19 (2004): 93-111.
7. Dennis Wheeler, "The Weather Diary of Margaret Mackenzie of Delvine (Perthshire): 1780–1805," *Scottish Geographical Magazine* 110, no. 3 (1994): 177–82.
8. "George Washington Carver and Tuskegee Weather Data," NOAA Central Library, accessed January 29, 2023, https://library.noaa.gov/Collections/Digital-Collections /George-Washington-Carver.
9. Fiona Clare Williamson, "POW Meteorology: Hong Kong, 1941–1945," last modified March 15, 2017, https://picturingmeteorology.com/home/2017/3/15/pow-meteorology -hong-kong-1941-45.

10. http://www.met-acre.org
11. Lucy Veale, Georgina Endfield, Sarah Davies, Neil Macdonald, Simon Naylor, Marie-Jeanne Royer, James Bowen, Richard Tyler-Jones, and Cerys Jones, "Dealing with the Deluge of Historical Weather Data: The Example of the TEMPEST Database," *Geo: Geography and Environment* 4 (2017): e00039.
12. Elizabeth R. Ellwood, Stanley A. Temple, Richard B. Primack, Nina L. Bradley, and Charles C. Davis, "Record-Breaking Early Flowering in the Eastern United States," *PLOS ONE* 8 (2013): e53788.

CHAPTER 5

Medical Science Archives

Closer and More Accessible Than They Might Appear

April White and David D. Vail

A Medical Science Past in Your City or Town

MEDICAL SCIENCE archives lay a solid foundation for interpreting science at history museums, historic sites, and historic houses. Often medical archives reflect some of the key forces that shaped communities, influenced local practices, and informed policies locally and more broadly afield. These materials support complex (and interesting) exhibitions and interpretive programs about the sociocultural, technological, and environmental contexts that affected health and wellness in hometowns, small communities, rural and urban America, then and now. Direct engagement with this two- and three-dimensional evidence can encourage local and out-of-town audiences to see themselves within the many facets of medical science.[1]

Medical science archives include highly personal as well as deeply revealing information about individuals, communities, authority, policies, and practices. They document relationships that played out between health practitioners and patients. They address the formation of the medical profession and the conflict between traditional practice and medical intervention. Careful analysis informs us about how local doctors worked in their communities and about how different approaches to medical treatment could affect relationships and lives. The development of the medical profession affords many opportunities to explore the theory and practice of science itself.

Finding sources to document medical science can start by contacting a local historical society. Collections often include personal papers of health practitioners including, for example, midwives. The Maine State Library holds Martha Ballard's diary, a sparse accounting of her daily routines as a midwife during the early national period on the Maine frontier. Historian Laurel Thatcher Ulrich based her award-winning *A Midwife's Tale* on this document. It and other histories of midwifery hold additional clues about the ways women built their understanding of human physiology and of relationships between bodies and the natural resources used to maintain women's health during their reproductive lives, and beyond.[2]

Historical societies often include personal papers, residences, or offices of doctors as part of their collections. Establishing parameters of use that protect privacy and ensure confidentiality can increase access to information about how doctors procured ingredients from local plants and processed these in keeping with centuries-old recipes in pharmacopeia. Doctors, pharmacists, nurses, and other health practitioners locally can help decipher meaning and methods from these historical sources and deepen the understanding of the science inherent in the historic practices.

Hospitals maintain corporate archives. These are generally closed to the public but initial contact can forge partnerships to tell timely stories. In contrast, county courthouses hold public records that can document policies and payments to maintain detention centers for populations who are older, incarcerated, or who have disabilities. State archives also hold records for mental health institutions and other comparable organizations. Volunteers, already knowledgeable about the records, can facilitate access and information retrieval, all within preestablished parameters to maintain privacy and confidentiality.

Careful analysis of medical science archives can help us reckon with uncomfortable history. For example, how does racism affect medical experiments? Who had authority to make decisions about treatments during the Tuskegee syphilis "experiments"? Who implemented forced sterilization of individuals in asylums and mental institutions? What moral and ethical dilemmas remain from race-based decision-making that affected Henrietta Lacks's treatments, but also led to the discovery of HeLa cells? What legacy remains of these decisions?[3]

A Case Study Bridging Town and Gown

In Nebraska's colleges and universities, medical science sources help connect the state's cities and town histories in important ways. Many campuses are known for distinct medical specialties and training programs, so their archival collections highlight those institutions, but their collections also show how local social views on health, wellness, and policies define and redefine communities. Engagement, then, requires us to think about interpretations that are accessible for all visitors—a medical science history for practitioners and the general public alike. And collaborations with private institutions such as Creighton University or Clarkson College help expand access to medical science archives for the public. For the University of Nebraska Medical College's (UNMC) Leon S. McGoogan Health Sciences Library, archival, artifact, and art collections related to medical science all present the dynamic history of both city and town throughout the state. In addition, UNMC has established the Wigton

Heritage Center, which makes numerous primary sources and objects accessible through a series of interpretive experiences that include physical as well as digital exhibits. A little farther west, the University of Nebraska at Kearney (UNK) campus joins this interpretive network with its historic role as the state tuberculosis hospital in the first three decades of the twentieth century.[4]

A Hospital That Became Part of Campus

UNK is fortunate to have the G. W. Frank Museum of History and Culture (Frank Museum) to tell this dynamic history of medicine and disease in the Great Plains. One of the most popular permanent exhibits, *The White Plague Comes to Kearney: A History of the Nebraska State Hospital for Tuberculosis*, features the Nebraska State Hospital for Tuberculosis (NSHT), which was the state's only hospital for tuberculosis patients. It operated between 1911 and 1972.

The White Plague Comes to Kearney presents primary source information from the Frank Museum's archives, alongside NSHT artifacts including the heavy black door from the medical waste incinerator, a wooden desk from one of the nurse's dormitories, a surgical gown worn by Dr. C. Van Dyck Basten, and items handcrafted by former patients in different decades. These artifacts help visitors connect with the past both physically and spatially. The entire curation comes together in a way that museum visitors find both contemplative and easy to absorb. This combined with friendly museum staff who are trained to present medical science–specific history help guide visitors comfortably through a subject that many may consider awkward, embarrassing, or even gruesome.

Along with the public exhibit, the Frank Museum continually looks to present medical science history that underscores larger sociocultural shifts. Just as medicine and medical treatments evolve to meet a population's needs, so too, do the methods for presenting histories in museums. As visitors desire to see more diverse and inclusive histories, the Frank Museum complements with new historical interpretations that reflect a more diverse past such as exploring the early twentieth-century career opportunities for women as nurses and hospital cooks and early instances of racial integration at the hospital. These revised interpretative perspectives in *The White Plague Comes to Kearney* exhibit offer powerful connections to various audiences but also serve a growing interest by scholars on the local history of disease and rural health care.[5] And the majority of the medical science artifacts and exhibits are in one of the former doctor's apartments, providing visitors a genuinely immersive historical experience. The Frank Museum also offers walking tours of the hospital grounds, complete with a map showing which buildings are still standing and used by UNK and which buildings no longer exist.

Another purpose of *The White Plague Comes to Kearney* exhibit is to dismantle older, harmful disease-related stereotypes and stigmas. For centuries, tuberculosis, commonly known in the early twentieth century as "consumption," was characterized by policymakers and newspapers in highly problematic racial and class-based terms as an outcome of poverty and immigration. *The White Plague Comes to Kearney* presents fact-based, responsible research that challenges these problematic depictions directly and educates visitors on their

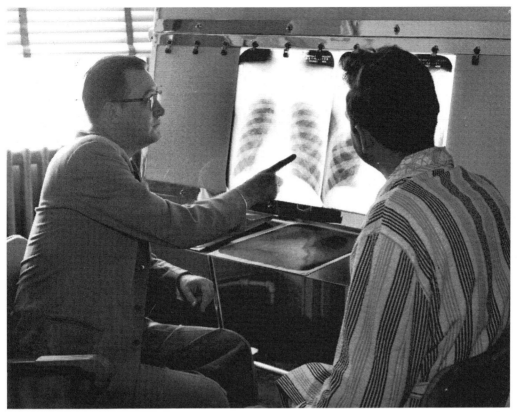

Figure 5.1. Dr. William E. Nutzman (left), Superintendent and Medical Director of the Nebraska State Hospital for Tuberculosis from 1950 until its closing in 1972, reviews a patient's (right) chest x-rays showing signs of tuberculosis of the lungs (pulmonary tuberculosis). *Credit*: G.W. Frank Museum of History and Culture, University of Nebraska at Kearney.

harm. Additionally, the Frank Museum's efforts to curate these medical science archives also helped many people see beyond the emotional shame and uncertainty that often accompanied a tuberculosis diagnosis in their family histories. Children of former tuberculosis patients who visit often voice their gratitude for this exhibit. Nearly everyone describes how their parents tried to hide their diagnosis so as not to worry brothers and sisters. Most also insist that keeping their illness a secret helped shield their families from being stereotyped as "poor" or "unclean" by local community members. After the children of former patients leave the Frank Museum, they often reflect on the comfort they feel and the renewed connections they have with their parents in a time of illness.[6] This goal, perhaps, is a good one for all visitors to *The White Plague Comes to Kearney* medical science exhibit—to blend the intellectual with the visceral to highlight a complex history of health, illness, and the local community.[7]

Moreover, curators of medical science archives can offer significant interpretive spaces for scholars, teachers, students, and public visitors. A curator can enter archival spaces and take an artifact's-eye view that connects the personal to the local in ways that are experiential and intellectual. The public can see how sickness, treatments, and health affected their communities in the past and how these relationships still matter now.

Notes

1. See Debra A. Reid, "Tangible Agricultural History: An Artifact's Eye View of the Field," *Agricultural History* 86, no. 3 (2012): 57–76. As Reid suggests, the basic goal in this approach is "to understand the objects as thoroughly as possible, and through that understanding, to gain a new perspective about a time period or a rural or farm process. There is not one fail-proof method. The key is to be aware as possible of the potential for increasing understanding by incorporating objects as historical evidence," 72.

2. Laurel Thatcher Ulrich, *A Midwife's Tale: The Life of Martha Ballard Based on Her Diary, 1785–1812* (New York: Alfred A. Knopf, 1990). See also Helen Varney and Joyce Beebe Thompson, *A History of Midwifery in the United States: The Midwife Said Fear Not* (New York: Springer, 2016); Judith Pence Rooks, *Midwifery and Childbirth in America: A History of Midwifery in the United States* (Philadelphia: Temple University Press, 1997); Lisa Yarger, *Lovie: The Story of a Southern Midwife and an Unlikely Friendship* (Chapel Hill: University of North Carolina Press, 2016); "The Historical Significance of Doulas and Midwives," the Smithsonian National Museum of African American History and Culture, accessed November 28, 2022, https://nmaahc.si.edu/explore/stories/historical-significance-doulas-and-midwives; and on women and public health, see "Women's History: Women in Public Health and Medicine," National Park Service, accessed November 28, 2022, https://www.nps.gov/subjects/womens history/women-in-health.htm.

3. See Ludmilla Jordanova, "Public History and the Public Understanding of Medicine: The Case of Embryology," *History Workshop Journal* 70 (Autumn 2010): 217–21; Michelle A. Hamilton and Rebecca Woods, "'A Wealth of Historical Interest:' The Medical Artifact Collection at the University of Western Ontario," *The Public Historian* 29 (Winter 2007): 77–91; Carolyn Birdsall, Manon Parry, and Viktoria Tkaczyk, "Listening to the Mind: Tracing the Auditory History of Mental Illness in Archives and Exhibitions," *The Public Historian* 37 (November 2015): 47–72; and Katherine Ott, "Disability and the Practice of Public History: An Introduction," *The Public Historian* 27 (Spring 2005): 9–24. For histories on the Tuskegee syphilis studies and Henrietta Lacks, see Susan M. Reverby, *Examining Tuskegee: The Infamous Syphilis Study* (Chapel Hill: University of North Carolina Press, 2009); and Rebecca Skloot, *The Immortal Life of Henrietta Lacks* (New York: Crown Publishing Group, 2011).

4. "Wigton Heritage Center," University of Nebraska Medical Center, accessed October 7, 2022, https://www.unmc.edu/wigton/; and "Leon S. McGoogan Health Sciences Library," University of Nebraska Medical Center, accessed October 7 2022, https://www.unmc.edu /library/special-collections/archives.html. For additional digital resources, see Jessica Ondusko, "Project Showcase: Exploring the Medical Heritage Library," *History@Work: The NCPH Blog*, last modified January 2, 2013, https://ncph.org/history-at-work/medical-heritage-library/.

5. Helen Kinnaman, interviewed by anonymous Frank House employee, G. W. Frank Museum of History and Culture Archival Collection, Kearney, NE, video recording, 2011; and Sadie B. Finch, "Fifth of Patients at State Hospital Here Are Children," *The Kearney Daily Hub*, October 1, 1921, 6.

6. Jerry Smith, interview by Dr. William F. Stoutamire, G. W. Frank Museum of History and Culture Archival Collection, University of Nebraska at Kearney, video recording, 2017.

7. Finch, "Fifth," 6.

SCIENCE AND THE HUMAN EXPERIENCE

Part II reveals how deeply science is embedded within the human experience. The contributors each focus on a topic (chickens, dahlias, tuberculosis, inadequate health care, irrigation and mountain snowpack, and food policy) to explore how we experience the world through science and the social impact of such interactions. Each also melds the past and present to present particularly compelling examples of how museums and historic sites can use everyday observations to link community, national, and international scientific outcomes. For Karen Scholthof, early twentieth-century narratives of backyard chickens and tuberculosis reveal the intimate connections we have with health and disease in our everyday lives. Our connections with the outdoors and contacts with domesticated animals oftentimes have deeper scientific meaning than realized at first glance. Emily Pawley also reflects on nature and science, using dahlia flowers that "traveled" from their native homeland in Mexico to become objects of beauty. An unstated aspect of the garden is how deeply it is imbued with science: flower shapes, colors, and their weediness offer jumping-off points to look at plant genetics, plant breeding, and the psychosocial benefits of experiencing nature.

Cherisse Jones-Branch shows us how strong actors in the Black community sustained the National Negro Health Week movement. They engaged with churches and federal public-health outreach to clean up neighborhoods and educate about venereal disease. By mid-century, funding for these segregated stratagems ended, folding into the federal programs. The takeaway message here is that science can be explored and understood from a platform of community outreach.

Both Patricia Rettig and Kristin Ahlberg bring us to how science develops locally and has national and international influences. Early landowners in arid locations, especially in the Western United States, often staked claims over water sources. This control limited access. Others sought adequate and fair distribution. Not until the engineering of a flume and measuring devices could legal water rights be monitored. The seminal tools and their deployment during the early to mid-1900s brings science to the fore, helping us become increasingly aware of drought and tensions between agricultural producers and urban citizens. Ahlberg challenges us to delve into public documents to understand the links between science and policy, using public laws to provide aid to underdeveloped countries with "Food for Peace" and "Food for Freedom" to save lives, but also to push US international interests and to support domestic agricultural and food manufacturers. Here, we can see how agriculture, public health, political science, nutrition, and economics all played a role and benefited by these actions. And, by extension we can take a fresh look at how government policy affects science, locally and globally.

CHAPTER 6

From Farm to Table

A One Health Scenario

Karen-Beth G. Scholthof

A GENTLE clucking sound emerges from the back of the garden as a dozen Rhode Island Red hens scratch for insects and prepare a dust bath. A couple of the hens are keen to follow you around in the yard, and occasionally are willing to snuggle. Their fresh eggs, with creamy orange yolks, are used for on-site cooking demonstrations and sold at the museum shop. These calm American-bred, egg-laying hens are important interactive, educational resources to explain the role of poultry at rural and urban sites. Yet, there is another dimension to the bucolic hen. Disease.

A visit to the grocery store brings this to the forefront. Every package of chicken has a label with safe handling instructions to keep raw meat refrigerated, to prevent cross-contamination of poultry and other foods, to cook the meat thoroughly, and to refrigerate leftovers promptly. Each component is intended to reduce the incidence of pathogens, particularly bacteria, causing disease in the consumer. To contextualize this, we have to look back more than a century to backyard chickens.

Chickens are extremely popular at house museums and historic sites.[1] The birds require little effort to raise, can free-range in the area, and help set the stage for curators who focus on life in the home and on the farm. They also are part of American history, with the Rhode Island Red one of dozens of breeds developed by homesteaders, commercial producers, and backyard poultry enthusiasts. There were "fancy fairs" and county fairs for egg and meat breeds; and chickens as subjects for entertainment, including in cartoons (Foghorn Leghorn and Chicken Little), children's books (*The Little Red Hen*), and dances (Chicken Dance polka).

The importance of healthy birds for egg and meat production, then and now, cannot be underestimated. For hens to earn their keep—mostly as egg layers for backyard chickens—good animal husbandry practices are essential. But there is another, often unstated, aspect that is equally important: reducing the potential for animal diseases infecting humans. Today's framework for this is One Health—an approach that recognizes that the health of humans, animals, and plants and their shared environment are connected and interdependent.[2] Ideas from the early 1900s, such as cleanliness and health, have been updated with One Health terminology: biosecurity, pathogens, environmental health, and food safety. Although "biosecurity" may have frightening overtones, its meaning is straightforward: taking mostly commonsense measures to prevent the introduction or spread of pathogens (and disease) in the flock; in turn, this will protect the human caretakers and consumers.

The primary focus of One Health is health, the meaning of which is both very broad and very vague when explaining its parameters. Keeping the focus on the backyard chicken, several science concepts can be used and developed by curators at house museums and historic sites. Animals bring a certain joy and excitement when viewed in these environs, a perfect setting to meld science with the humanities. One Health also provides a platform to discuss environmental sciences and how local decisions can affect the community at large. The Centers for Disease Control and Prevention (CDC) has One Health information on "how to achieve the best health outcomes for peoples, animals, plants and the environment" and an interactive disease-detective scenario, "Solve the Outbreak," to role-play these concepts.[3]

Figure 6.1.   Model poultry house and a member of the Lancaster Unit of Woman's National Farm and Garden Association, 1918. THF288962. From the Collections of The Henry Ford.

Backyard chickens require a certain level of care to keep them healthy and laying eggs. With increasing interest in raising backyard poultry, the US Department of Agriculture's (USDA) "Defend the Flock" program is an important resource to learn more about fowl health. This is in keeping with a long-standing mission of the USDA to protect livestock.[4] A century ago, the USDA distributed free documents, such as the *Farmers' Bulletins*, with poultry husbandry tips directed to farmers and enthusiasts in urban and rural settings.[5] During World War I, the USDA promoted raising chickens as an ideal strategy to convert food scraps and waste into food and an "economic opportunity for city families."[6] This push for domestic poultry and egg production was a component of government propaganda to ensure that beef and pork products would be available for the troops.

Chickens also were put to work as educational tools for elementary school children to improve their math, reading, and writing skills.[7] From this, One Health concepts can be used to interpret science within a historical context. In the early decades of the twentieth century, *Farmers' Bulletins* explained the importance of sanitation that included providing birds with access to sunny, well-drained areas and regularly removing poultry litter from the premises. Detailed instructions and drawings for coop designs in rural and urban settings showed how to keep the flock safe from rats and other predatory animals. Nutritional advice for raising poultry was straightforward, recommending food scraps from the table as a feed, but avoiding too salty foods or decayed meat, "as these products are especially injurious to fowls."[8] Fish scrap, milk products, and cottonseed meal were acceptable substitutes for meat; most vegetables needed no special preparation or cooking.

Beginning in 1918, mailing one-day-old chicks was permitted, a robust market that continues today. This added to a practice that began in the late-1800s, when eggs, dressed fowl, meats, and produce became a key part of the market economy, shipped by the US Postal Service. Some mail-order businesses catered to the bird fancier, others focused on expanding poultry production across the United States. This long-distance movement of chicks from farms to urban centers and homesteads created a new problem: Poultry diseases were on the upswing. When birds were hatched and raised at the same location, or with minimal transit, diseases would tend to remain localized. But like the birds, pathogens traveled. Pullorum disease, caused by the bacterium *Salmonella enterica* Pullorum, could cause up to 100 percent mortality in young chicks (two to three weeks old). Diseased mail-order chicks could easily infect birds distant from their origin site.[9]

Bacteriology was in full swing by the 1900s, and as early as 1913 the first serology tests were developed to detect salmonella in infected birds. Testing methods had progressed with the US government and state governments working together, and in 1934, by an act of Congress, the National Poultry Improvement Plan (NPIP) was established.

The NPIP linked health to cleanliness in 1935. As the NPIP explained: "Health is the foundation of successful husbandry" because production and reproduction both depend on it. And cleanliness ensured health, as the NPIP explained: "Constitutional vigor and sanitation provide the best insurance against ill health. . . . Since the spread of the majority of the more important poultry diseases results from conditions of filth, it becomes evident that several disease conditions are amenable to sanitary measures."[10] This advice to maintain sanitary living conditions to ensure healthy flocks remains foundational to the modern poultry industry.

Figure 6.2.   U.S. Postage Stamp issued in 1948 recognizing the centennial of the American Poultry Industry. From the Collections of the Smithsonian National Postal Museum.

Today Pullorum is not often seen, but the NPIP continues its flock health efforts, monitoring for bacterial and viral diseases, some of which also can "spill over" to humans and to wild birds.[11] This is the importance of purchasing chicks from established hatcheries that display credentials from the NPIP that birds are free of salmonella and other poultry pathogens. On October 14, 2022, the USDA announced a new program to reduce the incidence of salmonella in chicken and turkey.[12] However, other *Salmonella* species are annually associated with at least 310,000 human bacterial infections, resulting in 27,000 hospitalizations and 425 deaths. The annual health care cost associated with this foodborne human illness in the United States is $4.1 billion. One Health concepts and practices offer a plethora of avenues to interpret historic concerns about health, sanitation, and living conditions, in the context of current issues based on our continued interaction with animals at the farmyard and our homes.

Suggested Online Resources

American Poultry Association Standard Breeds, https://amerpoultryassn.com/poultry-breeds/.
Defend the Flock program, https://www.aphis.usda.gov/aphis/ourfocus/animalhealth/animal-disease-information/avian/defend-the-flock-program.
National Poultry Improvement Plan, US Department of Agriculture, https://www.aphis.usda.gov/aphis/ourfocus/animalhealth/nvap/NVAP-Reference-Guide/Poultry/National-Poultry-Improvement-Plan.
One Health, Centers for Disease Control and Prevention, https://www.cdc.gov/onehealth/.
One Health, US Department of Agriculture, https://www.usda.gov/topics/animals/one-health.
Partnership for Food Safety Education: Fight BAC! Curriculum Materials (K–12), https://www.fightbac.org/kidsfoodsafety/curricula-and-programs/.
[Robert] Frost on Chickens, a National Agricultural Library digital exhibit, https://www.nal.usda.gov/exhibits/ipd/frostonchickens/exhibits/show/henhouses.

Notes

1. Exploring One Health at a museum or historic site does not require poultry or livestock on-site. The attendant costs of animal husbandry, local laws, and ensuring the health and safety of museum visitors may make live animal exhibits an unrealistic goal. For guidance on this topic, see the "Livestock Care in Museums," *Association for Living History, Farm and Agricultural Museums*, February 8, 2017, https://alhfam.blog/2017/02/08/livestock-care-in -museums-resources/. Also, Jonathan D. Kuester and Debra A. Reid, "Livestock in Agricultural Interpretation," in *Interpreting Agriculture at Museums and Historic Sites* (Lanham, MD: Rowman & Littlefield, 2017), 187–201.

2. For a general introduction, see the Centers for Disease Control and Prevention: One Health, at https://www.cdc.gov/onehealth/.

3. The CDC is an authoritative resource for up-to-date One Health information, including "Health Topics: A to Z," a guide to diseases and microorganisms, available at https://www .cdc.gov/health-topics.html; *Salmonella* and food safety, https://www.cdc.gov/salmonella/; and "Solve the Outbreak," https://www.cdc.gov/digital-social-media-tools/mobile/applica tions/sto/web-app.html.

4. See the USDA's "Defend the Flock" program, updated January 7, 2022, https://www .aphis.usda.gov/aphis/ourfocus/animalhealth/animal-disease-information/avian /defend-the-flock-program/defend-the-flock-program.

5. The USDA *Farmers' Bulletin* series is archived in the University of North Texas Digital Library, accessed November 14, 2022, https://digital.library.unt.edu/explore/collections/USDAFB/.

6. *The Agricultural Situation for 1918. Part IX Poultry. One Hundred Hens on Every Farm—One Hundred Eggs from Every Hen* (Washington, DC: Government Printing Office, 1918), 12 pp.; and Harry M. Lamon, *Hints to Poultry Raisers, Farmers' Bulletin* 528 (Washington, DC: US Department of Agriculture, 1913, rev. 1918), 16 pp., https://archive.org/details /CAT85816930.

7. Franklin E. Heald, *Lessons on Poultry for Rural Schools*, Bulletin No. 464, States Relations Services (Washington, DC: US Department of Agriculture, December 1916), 34 pp., available at https://archive.org/details/lessonsonpoultry464heal/ (accessed November 5, 2022).

8. Rob R. Slocum and Alfred R. Lee, "Back-yard Poultry Keeping," *Farmers' Bulletin No. 1331* (Washington, DC: US Department of Agriculture, 1924), 18 pp., https://archive.org/details /CAT87203169/.

9. According to the *Merck Veterinary Manual*, Pullorum is uncommon today and is managed by good biosecurity measures. It is egg transmitted (transovarial), but primarily by direct contact with infected birds or indirectly via contaminated water, feces, or feed. For more information on diseases of poultry, including Pullorum, see the *Merck Veterinary Manual*, https://www .merckvetmanual.com.

10. *National Poultry Improvement Plan* No. 14, Animal Husbandry Division, Bureau of Animal Industry (Washington, DC: US Department of Agriculture, September 1935), 13 pp., available at https://archive.org/details/CAT30498734 (accessed October 14, 2022).

11. Beginning in April 2022, the poultry industry is weathering another crisis: the highly pathogenic avian influenza virus (HPAIV) outbreak of the H5N1 strain in wild and farm-raised birds. As of February 2023, H5N1 influenza A virus in flocks has resulted in the loss (due to death and culling) of fifty-eight million birds in the United States, resulting in record-high egg prices in the grocery store. This "bird flu" virus is evolving and can now infect mammals,

which is of extreme concern to virologists and to human and veterinary health practitioners. As of February 9, 2023, the USDA has identified H5N1 "bird flu" in mink, skunks, raccoons, red foxes, bobcat, grizzly bear, mountain lion, and marine mammals, https://www.aphis.usda .gov/aphis/ourfocus/animalhealth/animal-disease-information/avian/avian-influenza/hpai -2022/2022-hpai-mammals. These "spill-over" events, and the worrying potential of this virus to affect human health, are being monitored by national and international agencies, including the CDC, USDA, and World Health Organization.

12. In 2021, there were an estimated 532 million chickens and turkeys on hand and production of 111 billion eggs in the United States according to the USDA National Agricultural Statistics Service publication "Chickens and Eggs 2021 Summary," published February 2022, https:// usda.library.cornell.edu/concern/publications/1v53jw96n?locale=en.

CHAPTER 7

Seeing the Museum in the Garden

Using Living History Museums to Teach the History of Plant Introductions

Emily Pawley

WHEN WE'RE outside in the United States, we're always in a museum of species introductions. We walk on lawns full of dandelions and plantain brought by European settlers in the 1600s and hike in forests crowded with honeysuckle and kudzu brought during the rage for East Asian garden plants in the 1800s. We drive past alternating fields of corn, a Central American plant brought to North America by Indigenous peoples around 1000 CE, and soy, brought to the United States by East Asian immigrants in the late 1800s.[1] To ecologists, understanding these histories matters a great deal. Without them it is impossible to understand a community of plants, gauge its health, maintain, restore, or eradicate it. Without public understanding of such histories, we can't build support for meaningful conservation and restoration policy in wild landscapes or help gardeners and farmers participate thoughtfully in the continued circulation of species.

Despite its ubiquity, however, many people find this museum of introductions hard to see. In the 2010s, biologists and educators began to complain of the widespread tendency to see everything living and green as natural, undifferentiated, or even invisible.[2] While the recent increase in houseplant and gardening popularity has begun to reverse this tendency, plant and garden markets rarely reveal the ecological and historical meanings embedded in the plants they sell. Even for those gardeners who pay attention to distinctions between native and introduced species, the histories that produced these categories are often poorly examined. Ecologists themselves still wrestle with the violent histories and xenophobic judgments built into language like "colonization," "invasion," and "alien," and the concrete links between the ways that some plants and people have been marked as not belonging and treated with violence.[3]

Living history museums can help the public connect with these questions by making space for visitors to examine living plants as interpreted through different historical moments and coming from a broad range of introductions. During the periods memorialized by many living history museums, after all, most people worked on farms. Detailed knowledge of domesticated and wild plants—for ornament, food, brewing, fiber, dyeing, and medicine—was part of everyday life for almost everyone. Plant history therefore is part of the core mission of such museums. Moreover, drawing attention to these plants' global and Indigenous origins can help museums break free of the illusory fog of presumed whiteness in which visitors sometimes try to wrap them.[4]

My consulting work with Old Sturbridge Village (OSV), a living history museum in Sturbridge, Massachusetts, has informed my thinking about this. In particular, conversations with former and current Horticultural Coordinators Ruth DiBuono and Camryn Sarles, respectively, shared insights that helped to shape this chapter. Though it looks at first like a preserved space, OSV is a carefully constructed composite: forty historic structures reassembled on 240 acres of gardens, farms, and woods, designed to represent a typical New England village from 1838. About 250,000 visitors pass through these spaces every year.

While there is some traditional museum signage, visitors mostly learn about the buildings and the space through the work of skilled interpreters. While OSV interpreters wear period costumes and demonstrate daily tasks, they are not "in character" as historical people. Instead, they engage visitors in improvised conversations intended to connect them to larger historical themes and research. In depicting a period when plant introductions were rapid, much publicized, and often state supported, OSV can help us denaturalize the landscape of plant introductions.

Over decades, OSV has developed gardens and fields to represent the different social positions held by the occupants of the buildings. Native crops like corn, squash, and Jerusalem artichoke jostle with newcomers like turnips and onions in the kitchen garden of the middling Freeman Farm. Lofty ornamentals grace the Towne Garden, and what would then have been new varieties of apple crowd the orchard. Interpreting these living collections can be challenging. Many visitors have difficulty perceiving the differences between plants at first, and, once they learn about the use of some plants, sometimes have to be kept from poisoning themselves on others, for example on the monkshood that OSV interpreters grow safely out of reach. Unlike many other curators, OSV farm and garden interpreters have to literally keep their displays of artifacts alive, in the face of unpredictable seasons and pest and disease threats. This has implications for their training, which requires at least a full year. Interpreting or working on a farm in planting season does not prepare you to interpret it at harvesttime, nor does it prepare you for the range of crops that grow across the season.

Moreover, interpreters note that it is impossible to perfectly represent a historic landscape. Trees planted at the founding of the garden and in the surrounding woods are much larger than they would have been in 1838, when much of New England was raw and recently deforested. Charming old apple trees at OSV are too old; grafted varieties had been common only for a couple of decades. Modern deer fencing around younger trees would have been unnecessary, since deforestation and hunting had radically reduced deer populations. Even documented heirloom varieties are not artifacts but the lineal descendants of artifacts, which have diverged from their historical ancestors over generations. But it is in describing these differences that many of the most interesting conversations can occur. Modern deer levels threaten forest health as well as apple trees, a conversation that is easier to understand if we see those levels as a historical artifact, not a fact of nature.

To manage visitor unfamiliarity, and help interpreters focus, given the wealth of possible material, it can be helpful to focus on spaces and species that visitors know well and might encounter again outside the museum. The Towne Garden is one such space—it's a patterned ornamental garden, laid out with brick paths and lawns, with an arbor and a summerhouse, populated with mostly familiar plants, like the dahlias that serve as focal points in the late summer and fall.

Now that dahlias are available at any garden center, their novelty in the United States is blunted. But in 1838, interpreters explain, dahlias were still new and exciting, acting as much more unusual markers of status. Moreover, they had not lost the story of their Mexican origins. Though nineteenth-century "floriculturists" described dahlias as having been "weeds"

Figure 7.1.   Dahlia in the Towne Garden, Old Sturbridge Village. Photograph by Camryn Sarles. Courtesy of Old Sturbridge Village.

Figure 7.2.   Dahlia. Watercolor by Mary Altha Nims, Bennington, Vermont, ca. 1840. Gift of Richard Seymour Bayham. From the Collections of the Cleveland Art Museum. Creative Commons.   https://www.clevelandart.org /art/1934.135.

in Mexico, it now seems possible that they were grown for culinary and medicinal purposes, part of an elaborate Aztec gardening culture also responsible for modern US garden staples like marigolds, zinnias, and cosmos.[5]

Treated as historical artifacts, dahlias can show us the shifting meaning of plants, from food to ornament and even perhaps to weed. They remind us of imperial connections and the continuing role of Indigenous people of the Americas in developing parts of everyday life, and they also show us how the same networks of botanists who gave us many of our modern botanical concepts and species names were also responsible for the massive imperial reshuffling of species around the globe.

If we can learn from the dahlia, the showy centerpiece of the showy Towne Garden, we can also learn from the margins of that same garden. There, a suite of familiar weeds, like dandelion, plantain, chickweed, and clover, can almost inevitably be found. Weedy margins make valuable museum spaces precisely because they are so hard to get rid of—the manicured center of the garden preserves waves of fashion and intentionality, but the weedy margins show the wreckage of such waves. Dandelions and garlic mustard were culinary greens that escaped from the garden in the seventeenth and nineteenth centuries respectively. Plantain, an unintentional accompaniment, ran so far ahead of colonists that native peoples referred to it as Englishman's foot.[6] Margin-loving clover, spread enthusiastically by a network of clover seed mills in the 1830s, fixes its own nitrogen—it was brought to the Americas to add soil fertility to pasturage. The same feature lets it thrive in sterile landscapes like the packed soil at the edges of sidewalks.[7] Spreading among these European newcomers, *Chenopodium berlandieri*, a local lamb's-quarter, can help interpreters expand on precontact European changes. It's a crop that became a weed, a representative of the native eastern North American agricultural complex, cultivated by Indigenous peoples before the arrival of corn and beans from Central America.[8] These weeds can thus teach visitors both the deeper history of introduction and the ecology of disruption.

While OSV has a long history of plant selection and interpretation, at many museums the enormous potential for conversations about plants and their histories still remains untapped. Moreover, at all museums, this remarkable way of engaging visitors remains underfunded. Living history museums offer us an unparalleled pathway to help the public understand the cultural context of plant introductions that continue to shape and disrupt our gardens, our food systems, and our "wild" spaces.[9]

Notes

1. Alfred W. Crosby, *Ecological Imperialism: The Biological Expansion of Europe, 900–1900* (Cambridge: Cambridge University Press, 1986); and Christine M. DuBois, *The Story of Soy* (New York: Reaktion Books, 2018).
2. Mung Balding and Kathryn J. H. Williams, "Plant Blindness and the Implications for Plant Conservation," *Conservation Biology* 30 (2016): 1192–99.
3. Daniel Simberloff, "Confronting Introduced Species: A Form of Xenophobia?" *Biological Invasions* 5 (2003): 179–92; Christian A. Kull, "Critical Invasion Science: Weeds, Pests, and Aliens," in *The Palgrave Handbook of Critical Physical Geography*, eds. Rebecca Lave, Christine

Biermann, and Stuart N. Lane (Cham, Switzerland: Palgrave Macmillan, 2018), 249–72; and Jeannie N. Shinozuka, *Biotic Borders: Transpacific Plant and Insect Migration and the Rise of Anti-Asian Racism in America, 1890–1950* (Chicago: University of Chicago Press, 2022).

4. Courtney Fullilove, *The Profit of the Earth: The Global Seeds of American Agriculture* (Chicago: University of Chicago Press, 2017); Robin Wall Kimmerer, *Braiding Sweetgrass: Indigenous Wisdom, Scientific Knowledge, and the Teachings of Plants* (Minneapolis: Milkweed Editions, 2015).

5. "Dahlias," *New England Farmer* 11, no. 9 (1832): 78; J. Donald Hughes, "The European Biotic Invasion of Aztec Mexico," *Capitalism Nature Socialism* 11, no. 1 (2000): 105–12; Paul Avilés, "Seven Ways of Looking at a Mountain: Tetzcotizingo and the Aztec Garden Tradition," *Landscape Journal* 25, no. 2 (2006): 143–57; and Paul D. Sorensen, "The Dahlia: An Early History," *Arnoldia* 30, no. 4 (1970): 121–38.

6. Crosby, *Ecological Imperialism*, 161.

7. Mauro Ambrosoli, *The Wild and the Sown: Botany and Agriculture in Western Europe, 1350–1850* (Cambridge: Cambridge University Press, 1997); and Emily Pawley, *Nature of the Future: Agriculture, Science, and Capitalism in the Antebellum North* (Chicago: University of Chicago Press, 2020), 191.

8. Natalie G. Mueller, Gayle J. Fritz, Paul Patton, Stephen Carmody, and Elizabeth T. Horton, "Growing the Lost Crops of Eastern North America's Original Agricultural System," *Nature Plants* 3 (2017): 17092.

9. For more about the histories of species introduction, see Peter Coates, *American Perceptions of Immigrant and Invasive Species: Strangers on the Land* (Berkeley: University of California Press, 2006). For approaches on native plant and ecosystem restoration, see Tao Orion, *Beyond the War on Invasive Species: A Permaculture Approach to Ecosystem Restoration* (White River Junction, VT: Chelsea Green Publishing, 2015); and Douglas W. Tallamy, *Bringing Nature Home: How You Can Sustain Wildlife with Native Plants* (Portland, OR: Timber Press, 2007).

CHAPTER 8

The Outdoor Life

Seeking the "Cure" in New Mexico

Karen-Beth G. Scholthof

TUBERCULOSIS (TB) is, for most of us, an esoteric and unfamiliar disease of the past. During the early twentieth century, TB, or in the vernacular, consumption, was a terrible disease—literally consuming the person as the bacterium spread through the lungs and oftentimes other organs. The following focuses on TB cure facilities, not on the biology of the disease itself. Interpreting science is not just limited to a focus on laboratory or field research practices. Rather, the intent of this chapter is to encourage curators of house museums and historic sites to interrogate local architecture and photographs from the early twentieth century for evidence of nearby TB treatment facilities that represent the period between the development of germ theory (1880s) and efficacious antibiotic treatments (1960s). A close examination of everyday objects and sites can help explain both scientific advances and extant links to current events related to microbiology and pandemics.[1]

The "great outdoors" is part and parcel of the American experience, yet a century ago the notion of outdoor life had a very different connotation. This particular lifestyle was associated with tuberculosis and an attempt to obtain a "cure" by living in the high, dry climate of the American Southwest. Many patients could not afford to relocate or were too sick to travel. Therefore, they were treated at home or at a local sanatorium. (This was prior to the mid-twentieth-century development of antibiotics to treat the underlying bacterial infection.) The built environment that resulted, specifically, the architecture and infrastructure of care facilities, documents public health and environmental influences on it during the first decades of the twentieth century.

Before the availability of effective antibiotics for tuberculosis, treatment focused on exercise and fresh air. This "cure" was facilitated by the sanatorium movement, initiated in Germany ca. 1836. It was not until 1882 that Robert Koch, a German bacteriologist, determined that a microbe caused the TB disease. Koch demonstrated that the bacterium *Mycobacterium tuberculosis* could be isolated from an infected animal, cultured in the laboratory, and reintroduced to healthy animals to recapitulate the original disease. This experimental protocol, known as Koch's postulates, in effect represents the beginning of germ theory and proof that microorganisms could cause disease.[2]

In 1884 Edward L. Trudeau, a physician in Saranac Lake, New York, with TB disease, became the first American to confirm Koch's study. In 1888 Trudeau reported on the results of an experiment using three groups of rabbits: two separate groups were confined with poor diet and dark, enclosed living conditions; either infected with the *M. tuberculosis* or healthy. All of the infected rabbits died; the healthy rabbits were reported to be malnourished, but alive. The third group of rabbits were infected, then allowed to run free for four months "in sunshine and fresh air" on a predator-free island on the nearby Spitfire Lake. Of these five rabbits, one died within a month, and the remaining four were "cured." Trudeau proposed that the environment—healthful living conditions, rest and relaxation, fresh air, and plentiful food ameliorated the devastating pathogenic effects of an active TB infection for his patients.[3]

Importantly, Trudeau emphasized that TB was an infectious disease and not a hereditary condition, indicating that the patient was not to be blamed for infection and a cure could be offered to all. Soon thereafter Trudeau opened a "cure facility" in Saranac Lake. The purpose of the facility was twofold: treatment of the individual and isolation of TB patients from the community at large. Soon boardinghouses in the village itself were also built or modified to offer the "cure," each with verandas, screened porches, and balconies to ensure that the patients could spend many, many hours per day—in all weather—breathing fresh outdoor air. This unusual "cure" architecture is readily evident in Saranac Lake and other similar facilities across the country.[4]

The practices instituted by Trudeau spread throughout the United States, with rapid acceptance by the medical community. Yet, the vast majority of those with active tuberculosis disease were not at private or public facilities, but at home. Even so, in the early twentieth century in the United States, one out of 170 Americans was undergoing treatment in a sanatorium. In 1900 there were 34 facilities in the country with around 4,500 beds; by 1925, there were 536 sanatoriums with almost 675,000 beds. Most sanatoriums preferred patients in the early stages of disease, with inpatient treatment continuing for months or years. Increasingly, sanatoriums also provided therapeutic surgical interventions, such as artificial pneumothorax, to collapse and "rest" the affected lung.[5] Trudeau proclaimed that one-third of sanatorium patients became well, although mortality within a year of entering a sanatorium was greater than 25 percent, and most patients died within five years of their TB diagnosis.[6]

The US government also instituted measures to treat service members and veterans by repurposing military installations as cure facilities. In 1899, Fort Stanton, New Mexico, became the first sanatorium of the US Public Health and Marine Hospital Service to treat soldiers and sailors. In 1899, Fort Bayard, near Silver City, New Mexico (an established

Figure 8.1.   The Noyes Cottage in Saranac Lake. The original house was constructed in 1898, then remodeled as a cure cottage with various extensions and expansions of porches and dormers until it obtained its final structure in the mid-1920s (Gallos, Cure Cottages of Saranac Lake, 53–57). Photograph: Karen-Beth G. Scholthof, ca. 2011.

civilian destination for the "cure"), was transformed into the first US Army tuberculosis hospital with an in-house research program; in 1920 the property was transferred to the US Public Health Service to treat tubercular veterans.[7] In these locations, similar to civilian locations, the military used tents, cottages, and dormitories to house ambulatory patients. By 1924, Fort Stanton administrators boasted of ninety-six double-occupancy screened tent houses with "roller curtains to protect against storms" and "outfitted with two beds, stove, electric lights, and other necessary conveniences."[8]

This "outdoor life" with a focus on rest, nutritious foods, and clean air was considered best practice to arrest TB disease in the early twentieth century. Locations with a mountainous arid climate were considered especially salubrious, and the high desert of the US Southwest soon became a destination for the "cure." New Mexico provided a climate that "invites invalids in large numbers from other States," including those with tuberculosis and their attendant health problems, and that was "invigorating and conducive to outdoor life practically the year round."[9] As documented in *The Land of Sunshine*, there were "ample accommodations offered by tent cities, hotels, sanitaria," and "various hot springs . . . which are gaining prominence for their potent medicinal virtues."[10]

Touting these benefits, city and state boosters and real estate agents appealed to "lungers."The Albuquerque Chamber of Commerce proclaimed, "Why sacrifice health for wealth

Figure 8.2. Tuberculosis tent, Fort Stanton, New Mexico, ca. 1904 [Carrington (1904), 219]. "Living quarters occupied by consumptive medical officer, Ford Stanton Sanatorium, New Mexico. The tent has a wood floor, partial wood siding, and is outfitted with a stove, a desk, a bed, chairs, and a large mirror." National Library of Medicine, Images from the History of Medicine, available at https://collections.nlm.nih.gov/catalog/nlm:nlmuid-101395538-img.

when Albuquerque has both?"[11] The influx of Anglo residents into New Mexico factored into the territory's pursuit of statehood, finally granted in 1912. Yet, New Mexico remained "a frontier state of vast area" with a "sparse though mixed population, some of whom are under the jurisdiction of the Federal Government."[12] In 1920 the population of New Mexico was around 430,000, with racial demographics of 57 percent Spanish American and Mexican, 3 percent Indian, most of whom lived on reservations, and "a sprinkling" of African Americans.[13] The remainder were Anglos, and of this group, upward of 60 percent "came there originally for the health of [themselves or] some member of their family."[14] Some seekers of the cure experienced improved health, after which they remained in the Southwest, opening businesses, farming and ranching, and establishing schools and colleges. They became established members of these communities.

Moving across the country was no panacea for the acutely ill. Many new arrivals with TB had few or no financial resources or family ties. As early as 1909, an essay in *The Journal of the New Mexico Medical Society* noted: "For the sick, who has not the means to procure physical and mental rest, proper food and proper care, the last straw of going to New Mexico proves a sad failure." The fact lay in the increasing numbers of indigent and homeless persons in New Mexico towns and cities, many of whom died in remote, rural areas or even public streets.[15] Soon New Mexico had the second-highest rate of TB in the United

States, after Arizona, also a preferred destination for the cure. In 1925 the mortality in the United States from TB was 104 out of 100,000.[16] In Albuquerque, however, the mortality was tenfold greater, the majority of whom had arrived with the hope that the environment alone would cure their disease.

Of course, as these nonresidents interacted with the locals in New Mexico and Arizona, the incidence of TB disease spread to where it also became an alarming public health concern in the Native American and Hispanic population. During the 1920s and 1930s in New Mexico, 20 percent of school-aged children had TB, with rates often twice as high for Spanish Americans. Native Americans, living at distances from populated areas, were somewhat isolated from TB exposure early on, but they lacked access to public health and medical care. And their potential for exposure changed with the rapid influx of the Anglo population with active TB disease. And the rise of government boarding schools greatly contributed to the incidence of this disease. By 1927, approximately two thousand Native American patients were being treated in TB sanatoriums; the Indian Medical Service (of the US Public Health Service) reported that tuberculosis disease was the leading cause of mortality for American Indians, accounting for 25.4 percent of all deaths. In 1929, it was estimated that the mortality rate of Native Americans in the United States was four times that of the Anglo population.[17]

With increased public health measures, TB disease was slowly declining in the general population. The first chemotherapeutic agents became available in the early 1950s, followed by efficacious antibiotics in the early 1960s. In New Mexico, the success of implementing these measures was clear. In 1900, 194.4 TB deaths out of 100,000 were recorded; by 1940, the TB death rate was around 75 out of 100,000; and, in 1960, with the advent of very effective anti-TB drugs, the disease was increasingly under control with 8.1 deaths out of 100,000. In 2021, in the United States, 0.2 deaths out of 100,000 were reported, with less than 10,000 active (infectious) TB cases. Today, the worldwide incidence is around 10 million cases of active TB, resulting in 1.5 million deaths annually. Of significant international public health concern is the increase in multidrug resistant and extensively drug resistant *M. tuberculosis* strains, accounting for 500,000 cases per year that are refractory to standard treatments.[18]

In this chapter, I have used the architecture and infrastructure related to tuberculosis as a means to interpret the best practices in public health in the United States in the first decades of the twentieth century. Cultural artifacts, including tents and cottages; architectural additions to houses; and, schools and hospitals with large windows that could be opened for fresh outdoor air; provide a plethora of visual records of the "cure" during the early twentieth century. Germ theory had shown *M. tuberculosis*, the bacterium, was the causal agent of TB disease, yet the available intervention was by manipulating the lived environment. Some of these environmental practices were beneficial: increased air circulation, prohibiting spitting in public venues, reducing (airborne) contact between the well and unwell, all were (and are) good public health practices to reduce the spread of this respiratory disease in the community.

Interpreting the Science

By the early 1960s anti-TB drugs effectively treated tuberculosis infections, which rapidly shifted the sanatorium movement into an anachronism.[19] What had been a feared disease in the United States faded into memory. Facilities closed and their original purpose was mostly erased—some were repurposed for other medical uses, redeveloped as spas and resorts, or abandoned. Today, these buildings, archival materials, and photographs serve as potent reminders of how the natural environment and the built environment were used in an attempt to control and reduce the incidence of tuberculosis disease in the pre-antibiotic era.

Notes

1. A range of traditional archival material, published primary sources, and interpretive media exist to support TB research. For the Southwest, as an example, see Paul M. Carrington, "Further Observations on the Treatment of Tuberculosis at Fort Stanton, New Mexico," *Journal of the Association of Military Surgeons* 14 (1904): 207–34, https://archive.org/details/journalasso ciat03meetgoog/; Thomas Spees Carrington, *Tuberculosis Hospital and Sanatorium Construction*, third ed. (New York: National Association for the Study and Prevention of Tuberculosis, 1914), https://archive.org/details/tuberculosishosp00carrrich; Neil M. Goldberg, "Tuberculin Skin Testing," *Rocky Mountain Medical Journal* 65 (1968): 66–70; Philip P. Jacobs, *Sleeping and Sitting in the Open Air*, Pamphlet 101 (New York: National Association for the Study and Prevention of Tuberculosis, 1917), https://archive.org/details/sleepingandsitt00jacogoog; and Nancy Owen Lewis, "Chasing the Cure in New Mexico: Tuberculosis and the Quest for Health" (Santa Fe: Museum of New Mexico, 2016). For a history of TB, see René J. Dubos and Jean Dubos, *The White Plague: Tuberculosis, Man, and Society* (1952; repr., New Brunswick, NJ: Rutgers University Press, 1996). For an example of local history resources that incorporate TB history, see *Saranac Lake History*, https://www.historicsaranaclake.org. Other approaches to interpreting TB history include memoir, e.g., Betty MacDonald, *The Plague and I* (1949; repr., Seattle: University of Washington, 2016) and historical fiction, e.g., Andrea Barrett, *The Air We Breathe* (New York: W. W. Norton, 2008). Videos available through the National Institutes of Health's National Library of Medicine often address the ways the disease affected racial or ethnic groups, e.g., *Cloud in the Sky* (Hispanic Americans), 1940, National Tuberculosis Association, https://collections.nlm.nih.gov/catalog/nlm:nlmuid-8800947A -vid; *Another to Conquer* (Navajos, American Indian), 1941, National Tuberculosis Association, https://collections.nlm.nih.gov/catalog/nlm:nlmuid-8700185A-vid; and *The Inside Story* (Anglos), 1952, National Tuberculosis Association, https://collections.nlm.nih.gov/catalog /nlm:nlmuid-8800778A-vid.
2. Prior to the nineteenth-century work of Louis Pasteur and Robert Koch, the role of microorganisms as causal agents of disease had gained currency from work on plants and their pathogens. See Arthur Kelman and Paul D. Peterson, "Contributions of Plant Scientists to the Development of the Germ Theory of Disease," *Microbes and Infection* 4 (2002): 257–60.
3. E. L. Trudeau, "Environment in Its Relation to the Progress of Bacterial Invasion in Tuberculosis," *Transactions of the Annual Meeting of the American Climatological Association* 4 (1887): 131–36, https://pubmed.ncbi.nlm.nih.gov/21407353/.

4. Philip L. Gallos, *Cure Cottages of Saranac Lake: Architecture and History of a Pioneer Health Resort* (Saranac Lake, NY: Historic Saranac Lake, 1985); Katherine Ott, *Fevered Lives: Tuberculosis in American Culture since 1870* (Cambridge, MA: Harvard University Press, 1996); and Nancy Owen Lewis, *Chasing the Cure in New Mexico: Tuberculosis and the Quest for Health* (Santa Fe: Museum of New Mexico, 2016).

5. LeRoy S. Peters, "Changing Concepts of Tuberculosis during Twenty-Five Years," *Southwestern Medicine* 24 (1940): 46–48, https://archive.org/embed/southwesternmedi2419unse.

6. Trudeau (1887).

7. Joan M. Jensen, "Silver City Health Tourism in the Early Twentieth Century: A Case Study," *New Mexico Historical Review* 84, no. 3 (2009): 321–61, https://digitalrepository.unm.edu/nmhr/vol84/iss3/2; Carol R. Byerly, "'Good Tuberculosis Men': The Army Medical Department's Struggle with Tuberculosis," The Office of The Surgeon General, Borden Institute, US Army Medical Department Center and School Fort Sam Houston, Texas (Washington, DC: Government Printing Office, 2013), https://medcoe.army.mil/borden-tb-good-tb-men.

8. H. J. Warner, "Abstract of Annual Report of United States Marine Hospital No. 9, Fort Stanton, N. Mex.," *Public Health Reports* 39 (1924): 251–56, quote 252, https://www.jstor.org/stable/4577043.

9. J. W. Kerr, "Public Health Administration in New Mexico," *Public Health Reports* 33 (1918): 1975–95, quotes 1983 and 1977, https://www.jstor.org/stable/457493.

10. Max Frost and Paul A. F. Miller, compiler and eds., *The Land of Sunshine: A Handbook of the Resources, Products, Industries and Climate of New Mexico* (Santa Fe: New Mexico Bureau of Immigration, 1904), quote 5, https://archive.org/embed/landofsunshineha02newm.

11. Letter from the Albuquerque Chamber of Commerce [written by Aldo Leopold] to Leo Goossen, January 16, 1919. From the collections of The Henry Ford, gift of Gordon Eliot White, https://www.thehenryford.org/collections-and-research/digital-collections/artifact/524483.

12. Kerr (1918), 1983.

13. Kerr (1918), 1977.

14. Kerr (1918), 1977.

15. Anon, "Traffic in Philanthropy.—The Chicago Physician's Club.—The Valmora Ranch," *The Journal of the New Mexico Medical Society* 5 (1909): 57–59, quote 57, https://hdl.handle.net/2027/hvd.32044103070280.

16. Forrest E. Linder and Robert D. Grove, *Sixteenth Census of the United States: 1940. Vital Statistics Rates in the United States, 1900–1940*. US Bureau of the Census (Washington, DC: Government Printing Office, 1943), https://archive.org/details/sixteenthcensuso00unit.

17. M. C. Guthrie, "The Health of the American Indian," *Public Health Reports* 44 (1929): 945–57, https://www.jstor.org/stable/4579216.

18. The Centers for Disease Control and Prevention (CDC), "Tuberculosis (TB)," March 22, 2022, https://www.cdc.gov/tb/.

19. Globally, however, TB continues its devastating effects on humanity, especially the poor and those living in crowded, unhealthful environments. For more on the current world situation, see "Health Topics: Tuberculosis," World Health Organization, accessed February 11, 2023, https://www.who.int/health-topics/tuberculosis.

CHAPTER 9

"In the Interest of the Health Conservation of the American Negro"

National Negro Health Week, 1915–1951

Cherisse Jones-Branch

THE NATIONAL Negro Health Week movement, or NNHW, provides an important opportunity to construct programming in house museums that focuses on the intersection of Black health, science, activism, and racism during the twentieth century. The NNHW is a lesser-known part of a larger collaboration between scientists, reformers, and health practitioners to counteract the impoverished conditions in which many Americans lived in the early twentieth century. Black Americans implemented NNHW programming that addressed local needs. Reports on this work exist across the continent, waiting to be mined as evidence of historic agency and initiative in the face of opposition and oppression.

Established in 1915, the NNHW was the brainchild of Tuskegee Institute founder Booker T. Washington. It originated two years earlier under the auspices of the Negro Organization Society of Virginia, which initiated a periodic community cleanup campaign that included homes, lots, and fields all over the state. State and local health departments and voluntary organizations supported these efforts. NNHW later became a national program through Booker T. Washington and the National Negro Business League, which he founded in 1900 in Boston, Massachusetts.

Washington understood the benefits of the cooperative interracial health venture in Virginia to improve conditions in Black communities. He transplanted the idea to Tuskegee Institute. This suited the needs of many African Americans living in rural and urban spaces in the early twentieth century who were deeply impoverished. Because of the lack of economic opportunity, racism, and discrimination, they routinely resided in communities where health care was unavailable, sanitation was abysmal, and housing and food insecurity was rampant. This all caused disproportionately higher rates of illness and death. According to one source, in 1915, death rates were 20.2 per 1,000 African Americans as compared to 12.9 per 1,000 for white Americans.[1]

Washington knew well the devastating impact of poor health access among African Americans. In his last speech before the National Negro Business League in 1915 in Boston, Washington boldly spoke about the importance of improved outcomes among African Americans by "calling the attention of our people to the unnecessary high death rate among us, and in urging them to put into practice such rules of hygiene, sanitation, and cleanliness as will tend toward better health and longer life."[2] He proclaimed National Health Improvement Week, which later became National Negro Health Week, and was gratified by its nationwide observance because it changed "conditions about the home which have a tendency to promote disease."[3]

NNHW was a local, statewide, and national effort to increase awareness and address health disparities among African Americans who experienced disproportionately high mortality rates.[4] Such African American organizations as the National Medical Association, the National Association of Colored Graduate Nurses, the National Association of Colored Women's Club, and the National Negro Business League, among others, were important conduits for providing educational, hygiene, and clinical information in Black communities "for the general sanitary improvement of the community and for the health betterment of the individual, family, and the home."[5]

Typically observed during the week of Booker T. Washington's birthday, April 5th, NNHW community cleanup activities entailed painting and decorating, planting, insect and rodent extermination, and home repairs. Health education sessions were provided through lectures, sermons, radio talks, newspaper articles, and exhibits, in addition to the dissemination of health education materials. Additionally, state and local health professionals provided clinics and health demonstrations to African Americans.

During World War I, much emphasis was placed on cooperation with state and local health departments to provide better health education in Black communities. In Madison County, Alabama, in 1917, such cooperation yielded considerable benefits when rural sanitation demonstrations led to a decrease in Black deaths by 150 compared to the previous year. The Metropolitan Life Insurance Company reported a 9 percent reduction in the death rate among its African American policyholders. Black-owned insurance companies noted the NNHW's impact as well in reports to the National Negro Insurance Association (NNIA). Some of the companies affiliated with the NNIA further developed social service and health departments to assist their clients.[6]

Increased health care efforts during the World War I years additionally added to the NNHW's credibility. Focused attention on disease control and early prevention of deaths contributed to an ethos that supported daily health and sanitary practices.[7] Chief among the NNHW's health concerns during the war years was the prevention of venereal disease, which was considered "one of the world's greatest health problems." Although this was a national problem in rural and urban communities regardless of race, NNHW advocates worked closely with the US Public Health Service (USPHS) division of venereal diseases and the International Social Hygiene Board, established in 1918, to provide education about the devastating impact of gonorrhea and syphilis, and asserted that the "Nation must be made free from venereal diseases."[8]

By 1921 the USPHS began collaborating with the NNHW and published its journal, *NNHW Bulletin*. Also, during this year, US Surgeon General Hugh S. Cumming convened the first NNHW conference in Washington, D.C. The USPHS began producing the *Health Week* poster in 1927.[9] From 1932 until 1950, the USPHS supported the NNHW in collaboration with Tuskegee Institute, Howard University, the National Medical Association, and the National Negro Insurance Association by supplying staff, facilities, and materials to support its national health care initiative.[10]

Figure 9.1. National Negro Health Week poster published by the U.S. Public Health Service in cooperation with the National Negro Health Week Committee, Tuskegee Institute, 1929. The Tuskegee University Archives, Tuskegee University.

The NNHW movement additionally provided opportunities for African American medical professionals. In 1929 in Poughkeepsie, New York, Viola Avery, a graduate of Harlem Hospital in New York City, was hired as the city's first African American health nurse to "work among the colored persons of the city."[11] Avery's work in Poughkeepsie's Black community led to her being one of the Poughkeepsie Neighborhood Club's speakers for the NNHW observance in 1950.[12]

Nationally, the NNHW was led by Dr. Roscoe Conkling Brown, a graduate of the Howard University School of Dentistry.[13] A native of Washington, D.C., Brown practiced there and in Richmond, Virginia, until 1915 when he became an instructor at the Richmond Hospital Nurses Training School. He also organized the chemistry department at Virginia Union University. During World War I, Brown worked with the field service officer of the Surgeon General of the Army as a lecturer and consultant. After the war, he became the head of the USPHS's Office of Negro Health Work.[14]

Brown, like Black leaders around the country, actively promoted NNHW. To encourage participation, African Americans could vie for awards for such community honors as the "Activities Award" for local events and the "Progress Trophy" for three successive years of achievement. On occasion, a national award was donated, and the winner was highlighted in the NNHW magazine. In 1936, for instance, African Americans in Spartanburg County, South Carolina, won a trophy donated by the National Clean Up and Paint Up Campaign Bureau.[15] In Maryland, in 1938, five counties were awarded "Class A" certificates for their activities by the USPHS's National Negro Health Week Committee.[16]

NNHW supporters also sought to incentivize young people who were the future of African American communities. In 1939, the weeklong celebration included a poster contest and a special health pageant titled "The Physical Culture Girl of 1889 vs. the Health and Physical Education Girl of 1939." Black communities were provided with a text of the pageant and instructions for its organization by the NNHW Committee, which by then had been taken over by the US Public Health Service.[17]

During World War II, NNHW often employed patriotic language to encourage African Americans' involvement. In the 1942 issue of *The Future Outlook* newspaper, the twenty-eighth NNHW observance focused on community preparedness through personal hygiene by encouraging African Americans to adhere to the "Keep Well Daily Dozen" list from the US Public Health Service that included such directives as "Keep home and premises clean. Ventilate every room you occupy," "Always wash your hands before eating," "Have your doctor examine you carefully once a year," and "Consult your dentist at regular intervals." *The Future Outlook*, based in Greensboro, North Carolina, also used the NNHW to advertise for the "National Defense Program for Improvement of Community Health."[18] The newspaper utilized similarly patriotic language in 1943 when African Americans were urged to join President Franklin Roosevelt in "this fight on the Home Health front!" to help keep Americans strong by taking care of their health.[19]

Information about NNHW was disseminated from myriad sources in Black communities. None were more important, however, than the church pulpit and African American civic and fraternal organizations. When plans were underway for the thirty-sixth NNHW observance in Orlando, Florida, in 1950, local pastor R. H. Johnson initiated health education through the Good Neighbors, Inc., a community organization with radio and public programs, health poster contests, free health examinations, and what was described as a "mammoth memorial program honoring the memory of the late Booker T. Washington." All ministers, churches, clubs, civic and fraternal organizations, and local, state, county, and city health officials and workers were invited to participate. Another clergyman, Reverend Reddick, delivered the "National Negro Health Week Sermon" on the Orlando-based WDBO radio station.[20]

Although NNHW was immensely popular in African American communities, in 1951 federal security administrator Oscar R. Ewing announced the program's termination, arguing that it was "in keeping with the trend toward integration of all programs for the advancement of the people in the fundamentals of health, education, and welfare."[21] Despite the NNHW's official end, Black leaders such as Dr. Roscoe C. Brown and other USPHS personnel continued to provide consultation services to African American communities. The Office of Negro Health Work, which had been led by Dr. Brown, was renamed the "special programs branch" and operated as a clearinghouse for information on community and educational health programs for African Americans.[22]

African Americans continued to celebrate National Negro Health Week well into the next decade in both northern and southern, rural and urban communities with support from local Black organizations because they always understood it to be "in the interest of the health conservation of the American Negro and the Nation at large."[23] In Decatur, Illinois, in 1952, the Women's Progressive Club observed NNHW by hosting an American Red Cross home nursing teacher as a guest speaker.[24] Because of the myriad health conditions that plagued rural Black communities, African American leaders steadfastly utilized churches as the most practical gathering place to relay health information. Such was the case during NNHW in 1953 when Frances Gammill, the Mississippi County Tuberculosis Association executive secretary, spoke to a captive audience at the Bethel African Methodist Episcopal Church in Blytheville, Arkansas.[25]

For the time that it existed, National Negro Health Week was one of African Americans' boldest efforts to counteract health care inequality. Booker T. Washington's initiative augments the lesser-known history of African American health care activism as part of the "long civil rights movement."[26] NNHW specifically addressed the overwhelming negative health outcomes in African American communities that resulted from what scholar Alondra Nelson described as "the distinctly hazardous risks posed by segregated medical facilities, professions, societies, and schools; deficient or nonexistent health care services; medical treatment; and scientific racism."[27]

The NNHW's existence was, on its very face, an act of Black resistance, one that was supported by African American and white organizations, albeit for dissimilar reasons. The realization of better health outcomes in Black communities was a conscious political act. Healthy Black bodies better served the cause of African American political advocacy. The interracial and cooperative nature of NNHW was intentional because Booker T. Washington, like Black leaders nationally, understood the necessity of procuring white allies who had the best access to much-needed medical and financial resources. Indeed, NNHW supporters often cleverly utilized white fear of sick Black bodies to obtain support to improve health conditions in African American communities. Yet little change would have been realized without the labor of local and national Black organizations for whom African

American health was always a primary concern. It was at the micro level of operation within Black communities, most commonly in African American churches and educational facilities, where the impact of poor health access and the unfortunate lived experiences of racism and marginalization were most obvious.[28] It was also, however, within these spaces that the greatest opportunity for monumental change existed for health advocacy.

Utilizing the NNHW's significant history in house museums helps to unearth the extent of African Americans' determination to implement science and health reform initiatives to ameliorate the devastating conditions in their communities. Black leaders understood the significance of their health activism before the NNHW's advent and after its demise. They skillfully deployed this knowledge to empower African American communities to counteract poor health outcomes well into the twenty-first century.

Notes

1. Sandra Crouse Quinn and Stephen B. Thomas, "The National Negro Health Week, 1915–1951: A Descriptive Account," *Minority Health Today* 2, no. 3 (March/April 2001): 44.
2. Tracy Banks, "Health Matters 100 Years Later: The Legacy of Booker T. Washington's National Negro Health Week Campaign," *The Griot* 35, no. 1 (Spring 2016): 17–18.
3. Banks, "Health Matters 100 Years Later: The Legacy of Booker T. Washington's National Negro Health Week Campaign"; *National Negro Health Week, 17th Annual Observance* (Washington, DC: Government Printing Office, 1931), 11.
4. Roscoe C. Brown, "The National Negro Health Week Movement," *The Journal of Negro Education* 6, no. 3 (July 1937): 553.
5. Brown, "The National Negro Health Week Movement," 553.
6. *National Negro Health Week, April 1 to 7, 1923, The Ninth Annual Observance* (Washington, DC: Government Printing Office, 1923), iv.
7. *National Negro Health Week, April 1 to 7, 1923.*
8. *National Negro Health Week, April 1 to 7, 1923,* 10.
9. Quinn and Thomas, "The National Negro Health Week, 1915–1951," 46.
10. "Discontinue National Negro Health Week," *The Illinois Times* (Champaign, IL), April 9, 1951, 2.
11. "Mrs. Viola Avery New Health Nurse," *Poughkeepsie Eagle-News* (Poughkeepsie, NY), June 1, 1929, 1.
12. "Neighborhood Club Arranges Observance," *Poughkeepsie Journal* (Poughkeepsie, NY), April 2, 1950, 3a.
13. "Public Health Man to Conduct Panel," *The Gazette and Daily* (York, PA), March 27, 1952, 10.
14. "Public Health Man to Conduct Panel," 10.
15. Roscoe C. Brown, "The National Negro Health Week Movement," *The Journal of Negro Education* 6, no. 3 (July 1937): 557.
16. "Pocomoke Negro Health Club Receives Citation," *Worcester Democrat* (Pocomoke City, MD), December 30, 1938, 12.
17. "National Negro Health Week," *American Journal of Nursing* 40, no. 4 (April 1940): 440.
18. "National Negro Health Week," *The Future Outlook* (Greensboro, NC), April 11, 1942, 5.
19. "National Negro Health Week," *The Future Outlook* (Greensboro, NC), April 10, 1943, 6.
20. "36th Annual Health Week Set," *The Miami Times* (Miami, FL), April 1, 1950, 7.

21. "Discontinue National Negro Health Week," *The Illinois Times* (Champaign, IL), April 9, 1951, 2.

22. "Discontinue National Negro Health Week," 7.

23. *National Negro Health Week, 17th Annual Observance* (Washington, DC: Government Printing Office, 1931), 10.

24. "Progressive Club to Hear Red Cross Teacher," *The Decatur Daily Review* (Decatur, IL), April 13, 1952, 40.

25. "Shorter College Head to Speak at Church Here," *The Courier News* (Blytheville, AR), March 25, 1953, 19.

26. Jacquelyn Dowd Hall, "The Long Civil Rights Movement and the Political Uses of the Past," *Journal of American History* 91, no. 4 (March 2005): 1233–63.

27. Alondra Nelson, *The Black Panther Party and the Fight against Medical Discrimination* (Minneapolis: University of Minnesota Press, 2011), 24.

28. Kirk A. Johnson, "A Black Theological Response to Race-Based Medicine: Reconciliation in Minority Communities," *Journal of Religion and Health* 56, no. 3 (June 2017): 1098.

CHAPTER 10

Irrigation History

Moving Water from Colorado Mountain Peaks to the Fruited Plain

Patricia J. Rettig

SETTLERS ARRIVING in the American West in the 1800s recognized the productive potential of the region's good soils. They also saw the challenge of little natural rainfall. Adopting techniques of the Indigenous peoples as well as adapting those from their own homelands, earlier Spanish settlers had established acequia systems to irrigate their crops. Later Anglo-European settlers dug individual irrigation canals and eventually started organizations to cooperatively build bigger systems.[1]

As more people moved to the West, seeking opportunity and incentivized by US government programs, irrigated agriculture provided the essential food and fibers people needed, as well as incomes and the foundations of urban life. To improve practices, irrigation education emerged at the newly established land grant colleges in the West by the 1880s. The growth of irrigation and its challenges at the end of the nineteenth century boosted related research at academic and government institutions.

Professor and government researcher Ralph Parshall spent his life dedicated to the water measurement aspects of irrigation science. He got his start at Colorado Agricultural College (renamed Colorado State University (CSU) in 1957) in Fort Collins as a student, graduating in 1904, and spent most of the rest of his life working there and for the US Department of Agriculture's Soil Conservation Service. Because he worked for more than half a century at a time when scientific methods of irrigation were being increasingly investigated, and because many historical documents of his work survive, Parshall provides an appropriate case study to examine irrigation science history and how that could be interpreted at museums and historic sites, especially through archival materials.[2]

Farmers irrigate crops when rain proves insufficient or unreliable during the growing season. Irrigation is most typical in arid regions of the world where soils are good and water is available to divert from natural sources. Inherently multidisciplinary, irrigation science combines agricultural research on soils and crops with a host of water-related research, including chemistry, hydrology, hydraulics, and atmospheric science. Often conducted in an applied manner by civil or agricultural engineers, irrigation science helps people understand such concepts as amount of water to apply, the timing of that application, and methods and equipment to conduct the application efficiently.

The efficient use of water is a fundamental challenge arising from the arid conditions that cause the need for irrigation. In highly developed regions, demand for water for all the users or desired uses often outpaces the supply. Even with its comparatively low populations, by the early 1900s the West was already facing this challenge. Parshall recognized that if irrigators could more accurately measure and know more precisely how much water they were moving to their fields, they could take only their legal allocation and not more, enabling the resource to be used by additional appropriators in Colorado's water rights system.

Figure 10.1.   Ralph Parshall measuring water in Parshall flume with Soil Conservation Service vehicle in background, 1946. Colorado State University Libraries, Fort Collins, http://hdl.handle .net/10217/81620.

Parshall and his research team at the USDA, tasked with improving irrigation, investigated numerous devices and techniques that could make the practice more efficient. The device they developed over several years to measure the quantity of moving water in a stream or canal eventually was adopted around the world. Known as the Parshall flume, it has helped improve irrigation for a century.

Parshall was neither the first nor the last person at Colorado Agricultural College (CAC) to work on improving irrigation education, research, and application. He followed in the footsteps of the country's first irrigation engineering professor, Elwood Mead, who taught at the college in the mid-1880s while simultaneously working as the assistant state engineer. Mead moved on to Wyoming in 1888, serving as state engineer there until becoming head of USDA irrigation investigations in 1899. This role led to the establishment in 1911 of the Fort Collins branch, where Parshall became director in 1918 following nearly a decade of teaching at his alma mater.[3]

In collaboration with Colorado Agriculture College, Parshall established two research laboratories, one on campus and the other on the Cache la Poudre River, northwest of Fort Collins in Bellvue, for testing flume and other equipment prototypes. Parshall and his team also studied sand traps for irrigation canals, enabling debris to be caught and washed out of the system rather than causing clogs and maintenance issues for farmers. The group also worked on designing various irrigation recording instruments and meters, all with an interest in improving irrigation.

Details about the work of Parshall and his colleagues exist not only in the scientific reports they published, but also in materials they created in the course of their research that now survive in archives. In particular, the Water Resources Archive at the Colorado State University Libraries holds historically important materials from Parshall and his team. These provide evidence of their experiments through data, drawings, and photographs. Details of the collaborations survive in correspondence and drafts. Results and outcomes exist not only in publications but also in presentations, speeches, and vendor catalogs and brochures.

The Water Resources Archive, established in 2001, preserves historically important materials documenting Colorado water history, stemming from the university's legacy as a land grant institution long focused on water-related research and teaching. While the archive documents water issues across the state, including river compacts, water quality studies, and endangered species, a significant portion focuses on irrigation, as that has been an early and ongoing contribution of Coloradoans to society.

Though limited materials documenting irrigation science survive from the earliest days, one important set of images shows how the subject was taught in the late 1800s. Part of the Irrigation Photograph Collection, the set of more than four hundred lantern slides appears to have been compiled by CAC professor Louis Carpenter in the 1890s. A blend of locally made images and ones purchased from photography vendors, the 5x3-inch glass slides show irrigation-related scenes, equipment, and practices from Colorado, other states, and around the world. The images could be projected onto a wall for lectures to give students visual examples of different irrigation methods. Today, such images not only provide a look into the history of irrigation, but also how it was taught.

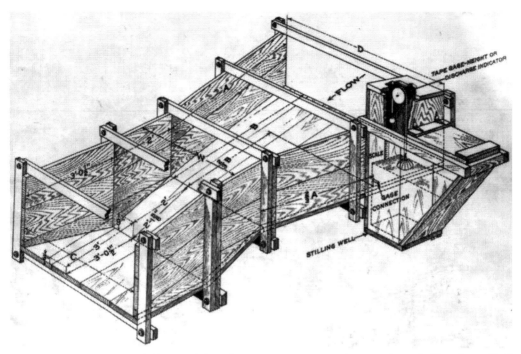

Figure 10.2. Drawing of flume. Colorado State University Libraries, Fort Collins, http://hdl
.handle.net/10217/68310.

Parshall likely learned from these images as a student. Though his own irrigation
research never included international locations, Parshall did extend his studies beyond
devices into two other significant areas. One was an investigation of the Colorado–Big
Thompson Project, a transmountain diversion proposed in the 1930s to bring water under
the Continental Divide to supplement irrigation on Colorado's eastern plains. The other was
snow studies, primarily for the Colorado River Basin, as a way to forecast how much water
would be in the river and available for a number of uses, including irrigation.

Parshall's documentation of this work ranges from extensive data, numerical analyses,
and reports to lighthearted poetry included in river water-forecasting conference meeting
minutes. Such variety shows some of the fun scientists were having amid very serious work.
When paired with construction photographs of the Colorado–Big Thompson Project or
snapshots of people taking mountain snowpack measurements, a fuller picture of the work
and the science behind it emerges.

As Parshall worked into the 1940s, irrigation continued to be taught and researched
at CAC by agricultural engineering and civil engineering professors. In 1947, Maury
Albertson had been hired to reinvigorate hydrology programs, not only through teaching
and research, but also through bringing in major grants and hosting national conferences.

Albertson was an enthusiastic professor and man of many ideas. Over several decades,
he taught various engineering classes, many of them related directly to irrigation structures
and systems or to the hydraulic principles essential for understanding and planning them.
He saved many of his class materials, including syllabi, assignments, exams, and lectures.
Known for taking classes on field trips to see dams, canals, and other irrigation infrastructure firsthand, Albertson also took photos at these sites and, as videotape technology

became popular, recorded trips as well. These videos, along with his own classroom lectures or those of guest speakers, offer another look into irrigation education.

Albertson also had a special interest in using science and engineering to improve life in developing countries. To that end, studies and reports about how to advance irrigation practices in places like India and Pakistan are present in his archival collection, giving an international view and providing a comparison with irrigation in the United States.

Amid his many interests, Albertson left the study of groundwater to his colleagues. A result of the Dust Bowl drought and improved technologies, pumping of groundwater for irrigation purposes emerged in Colorado by the 1940s. The addition of this water supply required increased scientific understanding of the underground flow, estimations of available quantities, and pumping techniques. At CSU, this in part manifested as engineers taking water-level measurements at observation wells along with scientifically developing a formula to simulate the effects on a stream of either the pumping or recharge of a well. The pioneering men who contributed to this, most prominently William Code and Robert Glover, left behind their field books, sheets of water-level measurement data, and reports explaining their findings.

Beyond the classroom or laboratory study of irrigation science, one can also examine the application of that knowledge. The Water Resources Archive holds the records of several irrigation organizations from across the state. Some of these collections solely focus on administrative materials like meeting minutes, annual reports, and water user information, but others include maps of irrigation systems, photographs, and oral histories.

While the Water Resources Archive is the only repository in Colorado that focuses on water history, other institutions hold materials relevant to irrigation science. Sibling archives at CSU, including the University Archives and the Agricultural and Natural Resources Archive, hold materials relevant to the administrative side of university research and teaching and to agriculture, respectively. The Agricultural and Natural Resources Archive holds records of CSU Extension and the Experiment Station, both important conveyors of irrigation science.

Other universities and some local history archives or historical societies hold materials more specific to their locations. The National Archives holds records of the USDA and its various units as well as other government agencies. Equipment companies and others that have adopted and adapted irrigation equipment and technologies may have their own archival holdings. These are just a few examples of where other important historical sources may be found. Local irrigation experts, including farmers, extension agents, and people in related professional organizations, also hold vast knowledge and are often generous in educating laypeople.

Ralph Parshall's Bellvue lab has languished over the decades after it went out of use. Now, though, a highly trafficked bike path runs past it, and a local group recently seized this unique opportunity to educate the public about irrigation history. The Poudre Heritage Alliance, a nonprofit organization that works to connect people with the abundant history in the federally designated Cache la Poudre River National Heritage Area, has begun restoring the site and creating interpretive signage. They will be using historic photos of the site and explaining the research that occurred there that has affected irrigation around the world. In addition to educating passersby, the site will raise awareness about the numerous resources on water history offered by both the Poudre Heritage Alliance and the Water Resources Archive.

Though using Parshall as a case study has focused on the research and education behind the engineering side of irrigation science, looking at subjects within crop science, soil science, atmospheric science, and even ecology should not be neglected.

Irrigation has been one way that humans have interacted with and adapted nature, especially through scientific understanding of crops, soils, and how to move water in a practice that ultimately feeds and clothes us. Conveying this history is increasingly important in a time when the western United States and places around the world that are heavily reliant on irrigation face severe and ongoing drought. Knowing how past people developed the science for the most efficient watering methods and devices can help those in the future better understand how to stretch the limited water supply.

Notes

1. Stanley G. Crawford, *Mayordomo: Chronicle of an Acequia in Northern New Mexico* (Albuquerque: University of New Mexico Press, 1988).
2. James E. Hansen II, *Democracy's College in the Centennial State: A History of Colorado State University* (Fort Collins: Colorado State University, 1977); Michael Weeks, "Measuring Expertise: Ralph Parshall and Watershed Management, 1920–1940," in *The Greater Plains* (Lincoln: University of Nebraska Press, 2021), 179–201.
3. James R. Kluger, *Turning on Water with a Shovel: The Career of Elwood Mead* (Albuquerque: University of New Mexico Press, 1992).

CHAPTER 11

Exporting the "Wonders of Modern Science"

Thinking Scientifically about Food Aid and Foreign Policy

Kristin L. Ahlberg[1]

WHEN RESEARCHING and writing, historians make numerous decisions in terms of what to emphasize. Following one line of inquiry, especially in the area of Cold War food policy, means that other motivations, questions, assumptions, and actors do not necessarily receive the attention they deserve. In researching the origins, implementation, and impact of Public Law 480, enacted in 1954 and commonly known as Food for Peace, I have previously focused on the program's use as a diplomatic tool that also served as one answer to the question of domestic commodity oversupply. Reframing one's research from a different perspective allows the historian to deepen their understanding and form more nuanced conclusions. Documenting the story of food aid during the 1960s by tracing the influence of science on commodity production and programming results in a more compelling history, one that reinforces arguments about the projection of US power during the midpoint of the last century. Examining primary sources such as public statements, government reports, and agency memoranda reveals the connections among science, agriculture, and foreign policy. Considering food aid from this perspective also provides more opportunities for reflection and discussion in these and related fields such as public history.

Highlighting the "food" in Food for Peace is an effort to elevate the importance of food and food studies as serious areas of inquiry on their own. Public historians Michelle Moon and Cathy Stanton assert that rather than just a stop along a longer path of interpretation, "food itself *contains* those questions [of interpretation] while offering an exceptionally accessible point of entry into them. We need to start with food and then *stay* there long enough to do that difficult work of unpacking its import."[2] In "unpacking" the confluence of science and food policy, I have chosen the Lyndon B. Johnson administration and its efforts to reauthorize and expand P.L. 480 because this era illustrates the extent to which food, and specifically nutritionally modified or newly developed commodities, played an important role in the conduct of a specific aspect of "soft power" diplomacy.

The archival and published primary sources reinforce the argument that Johnson intended to replicate his domestic Great Society reforms on a global scale. The sources reveal the assumption that commodities, and, in some cases the conditions placed upon them, would ease hunger, transform agricultural sectors, and generate new markets. For the purposes of this chapter, I consulted the *Public Papers of the Presidents of the United States* series; several volumes in the *Foreign Relations of the United States* series; P.L. 480 annual reports; White House Office of Food for Peace, Agency for International Development (AID) and US Department of Agriculture (USDA) newsletters; and memoranda from the Lyndon B. Johnson Presidential Library. Both Moon and Debra Reid, in their respective writings, advocate using these types of sources. Moon offers a caveat: "Of course, any sources must be read critically and considered alongside other contextual materials," while Reid suggests that one undertake research in the materials "to become more informed, both about evidence and details."[3]

The Food for Peace program neared its tenth anniversary at the beginning of the Johnson administration. Increased use of fertilizers, insecticides, and hybrid seed had resulted in postwar agricultural surpluses.[4] Intending to liquidate these surpluses and improve domestic prices, President Dwight D. Eisenhower signed P.L. 480 (the Agricultural Trade Development and Assistance Act of 1954) into law on July 10, 1954. As a result, aid, "which had often been a combination of ad hoc responses to famine or disaster abroad, became institutionalized under P.L. 480."[5] The legislation authorized the United States to sell commodities for foreign currencies to "friendly" nations, provide commodities to private voluntary and religious organizations for use in their feeding programs, contribute funds and supplies for domestic emergencies, and permit the barter of commodities for strategic materials.

While Eisenhower viewed P.L. 480 as a temporary program, President John F. Kennedy saw surplus crops as part of a broader rethinking of agricultural production within an expanded international development strategy. All forms of US foreign assistance would address inequities and strengthen the recipients' ties to US economic and political ideologies. This philosophy, and the intent to streamline foreign assistance, guided Kennedy and other officials in establishing AID in 1961. It also reflected Secretary of Agriculture Orville Freeman's belief that the United States could better "manage" agricultural abundance in order to close the hunger gap, while also exporting the technical "know-how" of agronomists to countries facing low agricultural yields and food insufficiency.

Such views reflected the importance development or modernization theory had on US policymakers during the 1960s. As articulated by Johnson administration official Walt Rostow and others, development theory presupposed that all countries moved through distinct stages of development. Offering commodities or other assistance, it followed, enabled

Figure 11.1. The U.S. Postal Service released the five-cent stamp, "Food for Peace— Freedom From Hunger," on opening day (June 4, 1963) of the World Food Congress (held June 4–18, 1963), in Washington, DC. From the collection of the Smithsonian National Postal Museum.

recipients to strengthen their own economic and political sectors and thus serve as bulwarks against communist expansion. "The modernizers," Michael Adas wrote, "assumed that all peoples not only could but would 'develop' along the scientific-industrial lines pioneered by the West."[6] Deborah Fitzgerald discerned a similar mentality in agriculture, with "mechanization and quantification in agriculture" resulting in a belief that successful, mechanized cultivation could be replicated everywhere.[7] Although scientific innovation had contributed to oversupply, science suggested a potential solution, as evidenced by AID administrator David Bell's testimony in February 1965 before the House Committee on Foreign Affairs. Exporting the expertise of the USDA, the land grant colleges and universities, and private industries, he asserted, would "help increase agricultural production in developing countries" as would the inputs "needed for raising agricultural output." In the interim, the United States would continue to make Food for Peace commodities available to nations facing food deficits.[8]

The continual focus on streamlining foreign aid and using food to advance US foreign policy objectives shaped Johnson's approach to P.L. 480. In remarks and speeches delivered in 1964 and 1965, the president underscored the connections between agriculture and foreign policy, characterizing food, as he proclaimed in his January 1964 State of the Union address, "as an instrument of peace."[9] In his February 1965 agriculture message, Johnson praised Food for Peace for it had "strengthened growing economies, contributed to rising standards of living, promoted international stability, and literally saved lives in many less developed countries."[10] Johnson, during an April 1965 address at Johns Hopkins University, linked science and agriculture to fighting the widening war in Vietnam. The United States would "expand and speed up" P.L. 480 shipments for nations throughout the Asian continent.[11]

The president envisaged a more wide-ranging foreign assistance program reflecting the objectives of his domestic War on Poverty, launched during 1964 and 1965. Foreign aid, including Food for Peace, had to address global illiteracy, hunger, and poor health. Unveiling the outline of these reforms during his January 1966 State of the Union address, Johnson exhorted Congress to "give a new and daring direction" to aid, "designed to make a maximum attack on hunger and disease and ignorance" in nations willing to undertake self-help initiatives.[12] In a separate foreign assistance message, Johnson explained that this was the cumulation of "the most sober review" and that he had directed several groups, including his General Advisory Committee (GAC) on Foreign Assistance Programs, to study every aspect of aid. "The incessant cycle of hunger, ignorance, and disease," he asserted, "is the common blight of the developing world. This vicious pattern can be broken. It must be broken if democracy is to survive."[13] He intended to rechristen P.L. 480, then facing its biennial reauthorization, as Food for Freedom, the first salvo in the global war against hunger. All agreements would be conditional upon recipients agreeing to undertake agricultural and economic improvements.[14] Although the administration failed to obtain the name change it sought, the resultant legislation, which Johnson signed into law in November 1966, included self-help provisions that, as the president explained, helped recipients' own "capacity to provide food for their people."[15] And, according to a late 1966 USDA publication, the P.L. 480 donation programs would now "stress" inclusion of "food for children that meet their requirements of proteins, minerals, and vitamins."[16]

The revamped program prioritized cooperation among government, land grant colleges and universities, and private industry in the continued development and production of enhanced foods, touted as the "wonders of modern science," including "high-quality synthetic foods as dietary supplements." The United States planned to encourage the production of similar foods within P.L. 480 recipient countries, as a condition of their aid agreements.[17] In testimony before the House Committee on Agriculture in support of Food for Freedom, Secretary of State Dean Rusk stressed that food aid remained an essential component of US foreign policy and that Food for Freedom embodied "two American traits—a compassion for human beings in need and a technical skill and productivity in agriculture which must be considered as one of the great success stories of modern history."[18]

From the early 1960s, the US agencies associated with administering P.L. 480 stressed the importance of nutrition and scientific innovation within an expanded food assistance program. The 1965 P.L. 480 annual report indicated that $2.5 million of AID-appropriated funds had been used to fortify nonfat dry milk, process flour and cornmeal with calcium, and add vitamins to other foodstuffs. As a result, "the program which for years has been principally concerned with combating hunger alone, is now geared to the elimination of the debilitating effects of vitamin and mineral deficiencies."[19] By 1967, the annual report touted the production within recipient nations of new products. In India, US funds allocated as part of a larger emergency aid program were used to manufacture balahar, a high-protein food.[20] The USDA's Economic Research Service (ERS), in a 1968 Agriculture Information Bulletin entitled "Science and Food for Freedom," was even more explicit in championing science in the fight against hunger, praising the importance of USDA research in producing higher crop yields, eradicating pests, improving nutritional qualities of foods, and recommending satellite surveying of fields.[21]

Not only did administration officials hail commodity fortification but they also endorsed cooperation between various government agencies as well as collaboration between the federal government and private industry in nutritious food development. The White House Office of Food for Peace reported that in 1964, Food for Peace director Richard Reuter convened a meeting of officials from "eight interested government agencies in an attempt to provide a closer link between the scientists and the program managers."[22] The P.L. 480 donation programs served as testing grounds for new fortified foods; the Soybean Council of America intended to test its vitamin-based soy flour beverage by including it in existing P.L. 480 Title II and III feeding programs.[23] AID also highlighted the public-private partnerships, noting several entities had collaborated with AID, USDA, and the National Institutes of Health in producing the new blended commodities CSM-Mix and Ceplapro.[24]

While altruism certainly influenced the desire to reduce world hunger, the scientific advances making this goal possible required new markets for sales and consumption to the benefit of US corporations. Freeman, referencing the collaborative effort expended to producing foods "from plants or from the ocean," believed that the United States "can look forward to the day soon when American food companies operating all over the world will be making a substantial impact of new protein foods in our battle against child and general malnutrition."[25] In 1967, National Security Council (NSC) staff member Howard Wriggins made the connections explicit, informing Rostow, who was at that point Johnson's National Security Adviser, that during a meeting with the staff of World Seeds, Inc., the seed purveyors discussed their success toward developing whye or triticale, a disease-resistant cross of wheat and rye that could be planted in food-deficit nations such as India. Wriggins conceded that the agriculturalists "were anxious for a commitment from the White House" that the US government "would finance large seed exports in the future."[26] That some of the largest food producers in the nation were involved in deploying new products under the guise of improved nutrition was inescapable, with Freeman drawing Rostow's attention to the fact that USDA had "encouraged the Coca Cola company, and now they are in Brazil with a soy protein drink—Saci."[27]

Foreign policy does not take place in a vacuum. Policymakers bring a variety of assumptions and beliefs to the table, and the culture within which they operate does impact the development and execution of policy. In this case study, I reframed a portion of my original analysis in order to develop a stronger argument about the importance of science to P.L. 480 and, by extension, to a specific component of US foreign policy. I encourage other scholars to revisit the archives or, in the case of public historians, their own collections to discover how focusing on the role of science can contribute to their narratives and public programming.

Notes

1. *The views expressed in this chapter are my own and not necessarily those of the Department of State or the US government. All sources cited are available to the public.*

    The Department of State applies unique editorial conventions when compiling and editing Foreign Relations, *some of which do not follow citation-style standards. For example,* Foreign

Relations, *edited by Carolyn B. Yee and Davis S. Patterson, includes four separate compilations: 1) Foreign Assistance Policy, 2) International Investment and Development Policy, 3) Economic Defense Policy, and 4) Commodities and Strategic Materials. Thus, the title is composed of two elements separated by a semicolon. Also, when citing to a* Foreign Relations *volume, the Department of State gives the description, then the volume and its title, followed by the document number.*

I would like to thank Laura R. Kolar, Kathleen B. Rasmussen, and Louise P. Woodroofe for their comments and suggestions. Many of the ideas I explore in this chapter stem from a session Debra Reid and David Vail organized for the 2017 National Council on Public History (NCPH) annual meeting entitled "Agricultural Policy and Science in Domestic and International Contexts: Public Memory, Public Interpretation."

2. Michelle Moon and Cathy Stanton, eds., *Public History and the Food Movement: Adding the Missing Ingredient* (New York and London: Routledge, 2018), 4. Emphasis is in the original; bracketed insertion is my own. See also Michelle Moon, *Interpreting Food at Museums and Historic Sites* (Lanham, MD: Rowman & Littlefield, 2016). For additional information about how historic sites incorporate food history in their public programming, see Megan Elias, "Summoning the Food Ghosts: Food History as Public History," and Adam Steinberg, "What We Talk About When We Talk About Food: Using Food to Teach History at the Tenement Museum," both *The Public Historian* 34, no. 2 (May 2012): 13–30 and 79–90, respectively.

3. Moon, *Interpreting Food at Museums and Historic Sites*, 176; Debra A. Reid, *Interpreting Agriculture at Museums and Historic Sites* (Lanham, MD: Rowman & Littlefield, 2017), 222–25.

4. Willard W. Cochrane, *The Curse of American Agricultural Abundance: A Sustainable Solution* (Lincoln and London: University of Nebraska Press, 2003), 9–10; Bruce L. Gardner, *American Agriculture in the Twentieth Century: How It Flourished and What It Cost* (Cambridge: Harvard University Press, 2006), 22, 24.

5. Willard W. Cochrane and Mary E. Ryan, *American Farm Policy, 1948–1973* (Minneapolis: University of Minnesota Press, 1976), 269.

6. Michael Adas, "Modernization Theory and the American Revival of the Scientific and Technological Standards of Social Achievement and Human Worth," in David C. Engerman, Nils Gilman, Mark H. Haefle, and Michael Latham, eds., *Staging Growth: Modernization, Development, and the Global Cold War* (Amherst and Boston: University of Massachusetts Press, 2003), 36. See also Michael E. Latham, *The Right Kind of Revolution: Modernization, Development, and U.S. Foreign Policy from the Cold War to the Present* (Ithaca and London: Cornell University Press, 2011), 57–58, 61.

7. Deborah Fitzgerald, *Every Farm a Factory: The Industrial Ideal in American Agriculture* (New Haven, CT: Yale University Press, 2010), 186–87.

8. "The Foreign Aid Program for 1966," AID administrator David Bell testimony before the House Committee on Foreign Affairs, February 4, 1965; Department of State *Bulletin*, March 8, 1965, 346.

9. "Annual Message to the Congress on the State of the Union," January 8, 1964; *Public Papers of the Presidents of the United States: Lyndon B. Johnson, 1963–1964*, Book I (Washington, DC: Government Printing Office, 1965), 117.

10. "Special Message to the Congress on Agriculture," February 4, 1965; *Public Papers of the Presidents of the United States: Lyndon B. Johnson, 1965*, Book I (Washington, DC: Government Printing Office, 1966), 147.

11. "Address at Johns Hopkins University: 'Peace without Conquest,'" April 7, 1965; *Public Papers: Johnson, 1965*, Book I, 397.

12. "Annual Message to the Congress on the State of the Union," January 12, 1966; *Public Papers: Johnson, 1966*, Book I (Washington, DC: Government Printing Office, 1967), 3.

13. "Special Message to the Congress on the Foreign Aid Program," February 1, 1966; ibid., 117, 118.

14. "Special Message to the Congress: Food for Freedom," February 10, 1966; ibid., 163–69.

15. "Statement by the President Upon Signing the Food for Peace Act of 1966," November 12, 1966; *Public Papers: Johnson, 1966*, Book II (Washington, DC: Government Printing Office, 1967), 1373–74.

16. "The New Food Aid Program" (Washington, DC: USDA, November 1966), 1.

17. "Special Message to the Congress: Food for Freedom," February 10, 1966; *Public Papers: Johnson, 1966*, Book I, 165.

18. "Food for Freedom Act of 1966," statement by Secretary of State Dean Rusk before the House Committee on Agriculture (read by Under Secretary of State Thomas Mann), February 25, 1966; Department of State *Bulletin*, March 28, 1966, 499.

19. *Food for Peace: Annual Report on Public Law 480* (Washington, DC: Government Printing Office, 1966), 14–15. See also memorandum from William Gaud to President Johnson, "Legislative and Administrative Accomplishments," October 1, 1965; *Foreign Relations*, 1964–1968, vol. IX, International Development and Economic Defense Policy; Commodities, Carolyn B. Yee and David S. Patterson, eds. (Washington, DC: Government Printing Office, 1997), Document 40 [the October 1, 1965, memo from Gaud to LBJ], and *The AID Story* (Washington, DC: AID, 1966), 18–19.

20. *Food for Freedom: New Emphasis on Self Help: 1967 Annual Report on Public Law 480* (Washington, DC: Government Printing Office, 1968), 28.

21. "Science and Food for Freedom," Agriculture Information Bulletin, No. 317, rev. version February 1968 (Washington, DC: USDA, 1968).

22. *Food for Peace: A Monthly Newsletter about Activities under PL 480*, Number 16, July–August 1964 (Washington, DC: Food for Peace Office, 1964), 5.

23. Ibid., Number 23, May–June 1965, 3.

24. "High-Protein Blended Foods—Newest Weapon in the 'War on Hunger,'" *Food for Freedom: A Monthly News Report on Activities Under PL 480*, Number 32, September 1966, (Washington, DC: Office of Material Resources, AID, 1966), 2–3.

25. Memorandum from Orville Freeman to President Johnson, April 15, 1966; folder PC 2 Food for Peace 11/6/65-8/18/66, box 4, White House Central Files, Peace, Lyndon B. Johnson Presidential Library, Austin, TX (hereafter cited as LBJL).

26. Memorandum from Howard Wriggins to Walt Rostow, "Meeting with Senator Burdick and Senior Staff of World Seeds, Inc.," February 17, 1967; folder CM/Wheat/8/13/65, box 13, White House Central Files, Commodities, LBJL.

27. Memorandum from Orville Freeman to Walt Rostow, April 19, 1968; folder 1968 Food Program, box 7, Subject Files, Federal Records, Agriculture, LBJL.

SCIENCE: A CULTURE OF DOUBT, A CULTURE OF QUESTIONING

The chapters in Part III examine the notion of "science as progress" to illuminate that it is possible to question and reflect on advances in science. This is in contrast to what is oftentimes the lay public's dismissal of facts and doubts about the veracity of science. This portends an important role for museums and historic sites—and visitors—to learn more about science and its process, toward asking and understanding questions about the role of science in our lives. With high public trust in museums, there is space to explore why the public oftentimes rejects scientific facts. What costs are paid by the individual and society when science research becomes the handmaiden of politics, when "controversial" science is defunded or banned and its findings questioned? And how can we place and explain science when there are difficult societal tensions to address, such as race, gender, class, health, and culture. Reflecting on "science as progress" brings in historical and contemporary complexities of genetic engineering foods, the palimpsest of atomic energy and nuclear weapons, white male hegemony and science training, confronting HIV/AIDS through the lens of race and sexuality, and our trust in science in our everyday lives, such as on the grocery shelf. And what is progress—is every advance worthy of our attention?

Debra Reid's chapter "Becoming a Scientist" calls attention to the lessons about race, class, gender, and authority that chemistry sets (and science kits in general) can teach. How did racism affect science training historically, in resources for training, support for individuals with science acumen, and access to scientific information? These seemingly innocuous educational toys reflected entrenched perceptions of a trusted chemist—a white male, specifically. Women and people of color did not see themselves in depictions of chemists sold with the sets; women appear only after 1947. The sets also cost money, putting them out of the financial reach of many. Overall, the sets did not inspire the public to think differently about who could be a scientist, thus perpetuating a narrow perception of who could be trusted in science.

Part III brings doubt and questioning to the fore: the difficult aspects of talking about science. Aimee Slaughter's experiences in interpreting "uncomfortable" science in Los Alamos, renowned as the location where scientists and engineers produced the first atomic bomb, led to the development of a reflection space for the visitor—one that allows for interpretation and thought about the deeper meaning of a science that transformed our world. The reflection space does not answer questions, but gives the visitor a chance to relive and meld experiences, and to consider that the culture of science is influenced by sociocultural experiences outside of the laboratory space. How will we citizens use our information and experiences to develop and direct our opinions to setting science policy? Jajuan Johnson delves into how one community—Black gay and same-gender-loving (SGL) males—addressed the burgeoning HIV/AIDS health care crisis beginning in the 1980s and extending to the present day. This community "unmarginalized" itself by becoming expert on the disease, social policy, and local and national organization. Groups formed to educate themselves and others to advocate for health care and to destigmatize the Black gay/SGL males who were nearly invisible due to a concatenation of issues—lack of access to health care, race, poverty, drug use, and stigma of their sexuality. Here the tensions of science and community loomed large: with the legacy of Tuskegee "experiments" and the real marginalization of Black gay/SGL men, how were they to trust science and medical care? Again, the answer came from within to learn the science, to strengthen the community through outreach (arts, education, public health, housing) and interpreting how science could best support their needs. Johnson's chapter reveals the many intersections of science and society that hinge on what it means to "trust" science.

Karen Scholthof and Debra Reid also bring up trust in science, which is further elaborated on by Benjamin Cohen. Scholthof and Reid show that the tools and methods of genetic engineering of foods are mostly a "black box," methods and outcomes that are unfamiliar to, and even indecipherable, for nonscientists, and thus a source of mistrust. As they discuss, the tools such as a "gene gun" were developed to solve a particular problem related to engineering plants for specific genetic traits—part of the scientific culture. Yet the public perceived the work as secretive or lacking in transparency—and this engendered doubt about the benefits of the technology. Cohen extends this by exploring food adulteration and state and federal efforts in the early twentieth century to ensure that the food supply was safe and to label foods. This analytic approach was transformative—chemistry (and later microbiology) became tools that the public accepted and depended on.

The tension addressed in this section prompts us to ask questions that tease out why in a time and place technology and science is accepted (or acceptable)? How do scientists come to understand that their experiments have meaning? Why do some groups accept science, while others view it as an artificial construct? To lighten this discussion, while making a serious point, Scholthof and Reid present tools found in the kitchen to challenge the notion that science only represents progress. They argue that the eggbeater did not represent an advance in kitchen tools when compared to the whisk—and yet hundreds of variants of the eggbeater were patented and could be found in every kitchen drawer in the twentieth century. The spork, they argue, is another device that fails to do the job—being not very useful for whisking eggs or for slurping soup. In all, this section shows the complexity of science and how it enters society. And that questions and questioning are the essence of "doing science."

CHAPTER 12

Reflecting on Uncomfortable Science

Aimee Slaughter

SCIENCE BELONGS in history museums because it's part of history: It is done by people, experienced by people, and part of our everyday lives. Because science is made by people, it is complicated, messy, and sometimes uncomfortable or even painful. The history of science is not stories of simple progress from one great idea to the next. The histories of science we share at our sites are human stories. These stories include joys and discoveries, new technological conveniences, and inspiration for the future; they also include exclusions and prejudices, extractive colonialism, violence, and war. Our historic sites need to make space for all these stories and the emotions they touch in ourselves and our visitors. Making this space is the work of connecting these histories of science to the core histories of our sites and providing space for visitors to reflect on what they have experienced at the site and in their own lives.

In 2015, the Los Alamos Historical Society took advantage of planned structural renovations to the historic building housing their Los Alamos History Museum and undertook a redesign of the outdated permanent exhibits. This small local museum presents histories of local and global importance: During World War II, the mountain community of Los Alamos was made into a secret Manhattan Project laboratory that developed the world's first atomic weapons. In this chapter I want to present the reflection space created in the updated museum as a small case study of the value of providing visitors a place to engage with uncomfortable histories of science. Doing this work helps provide our visitors a fuller picture of the past and creates avenues to make connections between the past, present, and future.

The Los Alamos History Museum interprets artifacts, photographs, and stories from centuries of local history, with a focus on the intense few years of the Manhattan Project (1943–1946). Most visitors are tourists to New Mexico, and if they arrive with any previous knowledge of Los Alamos history it is almost certainly about this wartime laboratory. The 2015 renovations maintained the museum's focus on the Manhattan Project, but among the important updates was the addition of a small reflection space where visitors could have a chance to both consider the uncomfortable history they have just encountered and to share their thoughts about it. Previously, there was not any space set aside in the museum for visitors to contemplate or process what they were experiencing. Visitors were often keen to discuss the Manhattan Project and the current implications of atomic weapons with docents, both before and after exploring the museum. Docents noticed that often visitors simply wanted to be heard. They also had questions, of course, but particularly when reflecting on how our atomic past has influenced the present, visitors often needed just to talk through their own thoughts with someone. The museum staff and the interpretation team considered these interaction patterns when redesigning the museum and responded to this visitor need by creating the new reflection space.[1] This intimate space provided visitors both a quiet place to reflect and an opportunity to share their thoughts on the connections between past and present in asynchronous dialogue with other visitors.

The visitors to our museums and historic sites are thinking about complex or even uncomfortable histories no matter how (or if) we interpret these stories. It can sometimes seem easier to avoid engaging with these histories because they are uncomfortable—perhaps especially in histories of science or technology, where the popular myth of straightforward, rational progress provides a comforting (though unrealistic) narrative. Anthropologist Martin Pfeiffer studies the nuclear past and present and has pointed out that visitors will notice when a historic site avoids engaging with a potentially difficult history.[2] He argues that nuclear heritage sites consistently downplay the suffering that nuclear weapons represent and only thinly mention influential antinuclear movements, and thereby do a disservice to visitors by failing to engage meaningfully with the full context of nuclear histories in their interpretation. He mentions the renovated Los Alamos History Museum in his analysis, and though I believe the updated Manhattan Project gallery and reflection space do begin to address uncomfortable nuclear histories, I agree there remains room for improvement, especially around antinuclear activism, as he specifically points out. Pfeiffer's article is an important reminder of how the interpretation we create for our visitors matters and how obfuscations or gaps can be read as erasures of history.

My use of the term "uncomfortable history" in this chapter is inspired by an online panel presentation from 2020, "Sitting Down with Uncomfortable Things in the Caltech Archives."[3] The presenters in this session reflected on archival objects and documents connected to the painful history of eugenics, and while they did not define their use of "uncomfortable," it is clearly in communication with the literature on difficult history. Julia Rose "broadly defines difficult histories as histories of oppression, violence, and trauma," and Magdalena Gross and Luke Terra further explain that difficult history is difficult to "the degree to which it challenges or undermines dominant societal narratives."[4] These scholars also emphasize that central to all difficult histories is their connection to our present world. (What history then, under close analysis, *isn't* difficult to some extent?[5]) It deserves mention that "difficult history" appears to be the accepted terminology in the literature, but the term itself is not ideal because it runs the risk of setting difficult histories in a separate category from "mainstream" or "normal" histories. Additionally, Gross and Terra remind us that "particular events are not 'difficult' for all students in the same way, nor do they stay difficult in the same way over time."[6] As we engage with difficult histories at our sites, it is important to keep in mind Julia Rose's reminder that our visitors are not the only ones learning—as history workers we are *also* learners, and in dialogues with visitors we can bring curiosity and empathy as we learn together.[7]

The primary difficult history interpreted by the Los Alamos History Museum is the history of the Manhattan Project. The museum itself is organized chronologically (or as close to chronologically as a historic-house-turned-museum allows): After exploring an orientation gallery giving an overview of the main eras of Los Alamos history, a temporary exhibition space, and the Ranch School gallery providing a picture of life in Los Alamos immediately before World War II, visitors enter the Manhattan Project gallery, which introduces the secret research lab and life in the wartime community. Off the end of this gallery is the small reflection space. Directly opposite the entrance to the space is a wall-mounted television and bench, and the wall on the left side has a small artifact case and interpretative panels. During renovations, the museum staff decided to make the storage room for the museum shop smaller—still functional, just demanding more efficient organization—in order to convert one end of the storage room into this new reflection space. The archway visitors walk through into the reflection space is effectively the threshold into the Atomic Age. Immediately before entering, visitors see photographs and artifacts from the weeks before the atomic bombings of Hiroshima and Nagasaki; inside the reflection space is interpretation and small, powerful artifacts about the consequences of these bombings, and with a button press the television plays a four-minute video introducing themes of this new era through oral histories and historic video footage. The lighting in the room is darker than in the rest of the museum to allow for viewing the video and to set a quieter, contemplative mood.

The reflection space's right wall is a black wooden pegboard covered with large and small clipboards. The clipboards display sheets with reflection questions, and a director's chair sits to one side in front of a small table with pencils and extra reflection sheets. This simple wall echoes the makeshift aesthetic seen in photographs from Manhattan Project displayed beside the clipboards, and despite its simplicity, it is truly central to how visitors experience the entire museum. Learning can happen best in uncertainty or discomfort, and a goal of this small space was to balance visitors' comfort levels in a zone that supports learning—making space for the uncomfortable history of nuclear war and its legacies while still creating a place where visitors felt welcome to share their thoughts.[8]

The reflection question on the clipboards matches the theme of the short film playing in the space, one of two options changed occasionally by museum staff to keep the exhibition fresh. Visitors are asked to reflect on either "What responsibility do scientists have for how their work is used?" or "What are citizens' responsibilities regarding government secrecy?" Both questions were intentionally crafted to inspire reflection on the Manhattan Project specifically or on society more broadly, allowing space for visitors to make connections between the past and the present.

Visitors engage with these reflection questions, usually reflecting on the two atomic bombings of Japan but sometimes responding on a more general level. Sometimes visitors respond directly to what previous visitors have shared. The questions are specific and not directly aligned with rote political talking points, so responses are rarely repetitions of current American political polarization but instead usually thoughtful, varied reflections. Some visitors share long-held beliefs while others seem to be working out their thoughts on paper as they respond, and many visitors reflect directly how this is a complicated history with no easy answers. Many visitors also make personal connections with history, using such phrases as "my father served," "if I were making the decision," "we all bear the shame," "I am a vet," "I am from Japan," or "I am a scientist."[9]

The wall of clipboards can also spark in-person conversations, and the reflection questions are available for guides as they engage with tours or field trips. Grappling with these reflection questions in person or on paper is not always easy—the history of the Manhattan Project is complicated and painful—but the museum team realized early in the redesign process that these conversations were a necessary and desired part of the museum experience. As with the International Coalition of Sites of Conscience's dialogic model, the goal was not to have visitors all agree on the same perspective when leaving the museum, but instead to have visitors all engage with the topic, recognize its importance, and feel their views were heard.[10]

In his work studying American nuclear museums, Bryan Taylor presents these museums as "reflexive spaces of 'entangled' rhetoric" where "visitors encounter a throbbing history marked by paradox and risk."[11] He finds that these museums usually try to impose "narrative 'containment'" on their difficult histories, which ultimately does not serve the needs of visitors and stakeholders, who are aware of the complicated connections between past and present even if the museum does not meaningfully engage with these legacies.[12] Museum spaces like the Los Alamos History Museum's reflection space begin to address his call for museums to "represent nuclear place and memory in new ways" rather than repeating familiar narratives that do not ask—and perhaps discourage—visitors to reflect on how history intertwines with our present lives.[13] According to the American Historical Association's recent nationwide survey on public perceptions of history, "most Americans agree that honest reckoning with their histories is needed, even if that history makes learners feel uncomfortable," and museum workers are recognizing that our visitors are ready to have conversations about uncomfortable histories at our sites.[14]

Atomic history has overlapped with the daily lives of Americans in many ways over the twentieth century. Some resources may be particularly relevant to the histories of your site, e.g., Matthew Lavine's *The First Atomic Age* explores public encounters with radiation in the half-century before the creation of the atomic bomb; *By The Bomb's Early Light* is a look at cultural responses to the atomic bomb written by Paul Boyer; and Spencer Weart traces the ways people have imagined, protested, and experienced atomic energy in *Nuclear Fear*.[15]

Atomic history, of course, is not the only example from the history of science relevant to historic sites, as the other contributors to this volume have shown. Nor is atomic history the only area where the history of science might become uncomfortable for learners. The history of science, technology, and medicine literature is far too large to try to summarize here, but the following books engage with uncomfortable episodes in American social history relevant to the missions of many historic sites. For example, in *Making Technology Masculine*, Ruth Oldenziel surveys the ways that technology, including consumer technologies, was imagined to be the realm of men despite contributions of women; in *For Her Own*

Good, Barbara Ehrenreich and Deirdre English investigate women's health, motherhood, and changing understandings of childhood over 150 years of history; and Ruth Schwartz Cowan provides a groundbreaking analysis of how American homes industrialized (and how new appliances and household technologies did not reduce the amount of time spent on housework) in *More Work for Mother*.[16]

A group of museum and collections professionals interested in the history of science convenes the Collections, Archives, Libraries, & Museums (CALM) caucus of the History of Science Society, which is an excellent resource for anyone interested in exploring how to connect science with their historic site.[17]

For the Los Alamos History Museum, the Manhattan Project is central to the historical narrative and clearly a difficult or uncomfortable history for many learners. Wartime histories—which broadly defined could even include stories from the Cold War—are one area of great overlap between histories of science, difficult histories, and histories important to the missions of many historic sites. What stories are visitors expecting when they visit your site? How might those histories be uncomfortable? How are the main stories of your site connected with histories of science, technology, and medicine? These histories are not stories of untroubled progress from one great discovery to the next—they are stories of people trying, failing, or succeeding to learn and change in painful, inspiring, unfair, uplifting, complicated, and messy ways. Science has changed the world in everyday and in era-defining ways, and these histories continue to affect our lives today. When these histories of science are uncomfortable, we have an opportunity at our historic sites to create space for our visitors to make their own meaningful connections between the past and present and reflect on how we should carve our future paths through the legacies of the past.[18]

Figure 12.1.   The reflection space of the Manhattan Project gallery in the Los Alamos History Museum in 2017. Photo by Kip Malone, courtesy Los Alamos Historical Society Photo Archive.

Notes

1. Before going too much further in this chapter, I want to acknowledge the team who worked on redesigning the museum and designing the new reflection space: a Los Alamos History Museum staff and volunteer interpretation committee, Quatrefoil Associates, Rainlake Productions, the Atomic Heritage Foundation for providing archival video footage, and the many local stakeholders who provided input individually and in group gatherings throughout the interpretation process.

2. Martin Pfeiffer, "Remembering an Incomplete Nuclear History," *Outrider*, July 8, 2020, https://outrider.org/nuclear-weapons/articles/remembering-incomplete-nuclear-history.

3. "Sitting Down with Uncomfortable Things in the Caltech Archives," Caltech Division of the Humanities and Social Sciences, October 2, 2020, https://www.youtube.com/watch?v=iHF6JEOLFVU.

4. Julia Rose, *Interpreting Difficult History at Museums and Historic Sites* (Lanham, MD: Rowman & Littlefield, 2016), 28; and Magdalena H. Gross and Luke Terra, eds., *Teaching and Learning the Difficult Past: Comparative Perspectives* (New York: Routledge, 2019), 5. Gross and Terra also detail five criteria for identifying difficult histories in their introduction to this volume and in Magdalena H. Gross and Luke Terra, "What Makes Difficult History Difficult?" *Phi Beta Kappa* 99, no. 8 (2018): 51–56, https://kappanonline.org/gross-what-makes-difficult-history-difficult.

5. This is also the conclusion reached by Bruce Vansledright and Sebastian Burkholdt in their chapter "Warts, Polyps, Blisters, and All? Problems in Learning to Teach a Provocative Past in a Troubling Way," in *Teaching and Learning the Difficult Past* (New York: Routledge, 2018), 119–36.

6. Gross and Terra, *Teaching and Learning the Difficult Past*, 5.

7. Rose, *Interpreting Difficult History at Museums and Historic Sites*, 29.

8. For more on learning in discomfort, see, for example, "Embracing Discomfort Can Open Our Minds to New Ideas," *Observer* 35, no. 4 (July/August 2022), https://www.psychologicalscience.org/publications/observer/obsonline/2022-may-embracing-discomfort.html; and Ephrat Livni, "A New Study from Yale Scientists Shows How Uncertainty Helps Us Learn," *Quartz* (July 31, 2018), https://qz.com/1343503/a-new-study-from-yale-scientists-shows-how-uncertainty-helps-us-learn/.

9. A collection of these visitor responses is available in the Los Alamos Historical Society Archives.

10. For examples of this dialogic model, see "Front Page Dialogues," International Coalition of Sites of Conscience, https://www.sitesofconscience.org/en/resources/frontpagedialogues/.

11. Bryan C. Taylor, "Radioactive History: Rhetoric, Memory, and Place in the Post–Cold War Nuclear Museum," in *Public Memory: The Rhetoric of Museums and Memorials*, Greg Dickinson, Carole Blair, and Brian L. Ott, eds. (Tuscaloosa: University of Alabama Press, 2010), 64, 57.

12. Taylor, 57.

13. Taylor, 76.

14. Section 9, "What Are the Public's Attitudes toward a Changing and Uncomfortable Past?" in *History, the Past, and Public Culture: Results from a National Survey*, Peter Burkholder and Dana Schaffer, eds. (Washington, DC: American Historical Association, 2021), https://www.historians.org/history-culture-survey.

15. Matthew Lavine, *The First Atomic Age: Scientists, Radiations, and the American Public, 1895–1945* (New York: Palgrave Macmillan, 2013); Paul S. Boyer, *By the Bomb's Early Light: American Thought and Culture at the Dawn of the Atomic Age* (Chapel Hill: University of North Carolina Press, 1994); and Spencer Weart, *Nuclear Fear: A History of Images* (Cambridge, MA: Harvard University Press, 1988).

16. Ruth Oldenziel, *Making Technology Masculine: Men, Women and Modern Machines in America, 1870–1945* (Amsterdam: Amsterdam University Press, 1999); Barbara Ehrenreich and Deirdre English, *For Her Own Good: 150 Years of the Experts' Advice to Women* (New York: Anchor Books, Doubleday, 1978); and Ruth Schwartz Cowan, *More Work for Mother: The Ironies of Household Technology from the Open Hearth to the Microwave* (New York: Basic Books, 1983).

17. To connect with the CALM caucus, visit their site on the History of Science Society's Web page: https://hssonline.org/group/CALM.

18. I gratefully acknowledge the support of an Emanuel Fellowship from the Consortium for History of Science, Technology, and Medicine in writing this chapter. I want to thank the colleagues who continue to help shape my thinking on this topic, especially the remarkably thoughtful panelists of our 2021 Society for the History of Technology/History of Science Society roundtable presentation Lara Freidenfelds, Johannes-Geert Hagmann, Kate Jirik, Kara Swanson, and Alex Wellerstein; Samuel Buelow; Stephanie Yeamans; and Nicholas Lewis. Thanks also to Rebecca Collinsworth and the staff of the Los Alamos History Museum, and to the editors for bringing this volume together.

CHAPTER 13

For Our Own Protection

On Black Gay Males and HIV/AIDS Activism

Jajuan S. Johnson

BLACK GAY and same-gender-loving males (SGL) historically and currently contribute to medical and scientific advancements in the fight against HIV/AIDS. In part, as a result of Black gay/SGL males building wellness networks, organizing politically, and crafting public policy expanding care access to citizens grappling with social determinants of health, today, HIV/AIDS is a manageable chronic condition. On the scientific front, Black gay/SGL males work in diagnostic labs, as care providers, and some guide public health policies to forward interventive measures. The testimonies and narratives of Black gay/SGL males, who are often shaded in the long story and harsh realities of HIV/AIDS, are crucial for the historical memorialization of the epidemic, particularly in the 1980s and 1990s.

Despite the piercing impact of HIV/AIDS on the lives of Black gay/SGL males, their voices are often muted on all fronts of the current realities of the virus and its history: "From the beginning of the epidemic, black gayness was often implied in numerous reports and stories, but black gay men remained largely hidden from view."[1] This chapter invites museum and public history practitioners to develop interpretation centralizing the resilience and struggles of a population of triply marginalized males who persist in the fight to end HIV/AIDS globally.

A Brief History of HIV/AIDS

The story of HIV/AIDS in the United States often starts with the 1981 outbreak that was primarily linked to its grim impact on white gay men diagnosed with *Pneumocystis carinii* pneumonia (PCP) and other life-threatening illnesses like Kaposi sarcoma (KS).[2] Before the identifier Acquired Immune Deficiency Syndrome (AIDS) emerged, the mysterious virus was referred to in stigmatizing terms such as "Gay Related Immune Deficiency Syndrome." "The perception of AIDS has been that of a disease that concerns 'others' and has been tainted with moral judgment and condemnation."[3] The stigmatization of people living with HIV/AIDS, particularly for people who identified as gay or lesbian, stymied awareness among heterosexual men and women and Black and brown people, with significant adverse public health consequences. In the seminal text, *Boundaries of Blackness: AIDS and the Breakdown of Black Politics*, scholar Cathy Cohen describes types of isolation Black gay/SGL males endured in the early era of the disease: "The first stories on AIDS in 1981 and 1982 paid little attention to its impact on African American gay men, African American injection drug users, or other members of African American communities."[4] In hindsight, the early assertion that the virus affected only white males bespeaks the ways African Americans, mainly, Black gay/SGL males, are made invisible in the health care system.[5]

The progression of HIV to AIDS occurs over years; thus, the virus was likely spreading within marginalized communities well before it was evidenced among white gay men in the United States. In this chapter, the narrative of HIV/AIDS starts with Robert Rayford, a sixteen-year-old Black boy who some scientists believe died of an early strain of HIV/AIDS on May 15, 1969, in St. Louis, Missouri.[6] The story of Rayford is littered with mysteries possibly tied to his identity as a Black male. Furthermore, he was poor, perhaps living with a mental disability, and a possible victim of sexual abuse. He also resided in the inner city of St. Louis during the 1960s, a tumultuous period of protest to upend racial discrimination. A Black boy died of an illness akin to AIDS; few people cared to know his story for decades, save a few journalists.

Rayford's case underscores the deep inequities in the health care system related particularly to African American males before the HIV/AIDS crisis of the 1980s and beyond. A June 2, 1995, CDC Morbidity and Mortality Weekly Report, "Update: Trends in AIDS Among Men Who Have Sex with Men, United States, 1989–1994," indicated an astronomical rise of HIV cases among Black and Hispanic men who have sex with men (MSM) while rates among white males declined. The disproportionate rates were attributed to the decreased access to HIV prevention services and culturally inappropriate HIV preventive activities. This reality illuminated the need for community planning to assess needs specific to Black and Hispanic MSM. However, data since 1994 related to Black gay/SGL males is still grim compared to white MSM. A December 2021 CDC report on HIV and African American gay and bisexual males indicated in 2019, 26 percent of new cases were among Black MSM, and about three out of four Black MSM who received an HIV diagnosis were between the ages of thirteen to thirty-four. The delay in linkage to care, inaccessibility to prophylactic and antiretroviral drugs, poverty, racism, stigma, and underrepresentation in biomedical research are enduring challenges.[7]

Silence surrounded Rayford's death, much like many Black gay/SGL males who died of HIV/AIDS decades after his passing. His narrative is instructive and can be used to peer closer into the historic maligning of Black gay/SGL males living with HIV/AIDS. Rayford's case situates the ways the lives of Black men and boys are obscured in the history of a medical crisis that still disproportionately impacts the lives of Black gay/SGL males and, more broadly, men who have sex with men.

Black Men Loving Black Men Is the Revolutionary Act

Since the earliest fight against HIV/AIDS, Black gay/SGL males have been on the local, national, and global front lines demanding health equity for themselves and people who do not identify as Black or LGBTQ+. The emancipatory vision of Black gay/SGL males is often inclusive, and their early activist work saved generations. Of course, the revolution started within. During the apex of the HIV/AIDS crisis, the message of self and communal empowerment was central in the cultural productions of filmmaker Marlon Riggs and writers such as Joseph Beam, Craig G. Harris, Melvin Dixon, Assotto Saint, and countless other artists who were casualties of HIV/AIDS. They provided a canon of literature, poetry, and performances that are primary sources ensuring the evidence of their earthly presence would not be tucked away in the folds of history. When their brothers fell, they continued fighting for medical advancements they would likely not experience.

It is essential to centralize their collective agency in presenting more authentic narratives of HIV/AIDS related to Black gay/SGL males. Despite the effects of HIV/AIDS in the early years, they engaged in radical placemaking critical to their health and wellness. Plainly stated, they formed communities encompassing the love, security, and autonomy they were often denied in heteronormative spaces. A nurturing community is essential for HIV/AIDS prevention and care, especially during a public health crisis disproportionately affecting an already highly vulnerable group of Black males. With few resources, they created community spaces to mobilize on individual and collective issues. The topic of this chapter derives from a poem by the late Black gay/SGL poet Essex Hemphill entitled "For My Own Protection," in his book *Ceremonies*. The opening verse, "I want to start an organization to save my life," encapsulates the intensity of the moment where Essex is probably grappling with the lingering HIV/AIDS crisis of the 1990s that later claims his life.[8] Hemphill declared that Black men are invaluable and that their salvation is linked to communities galvanizing to dismantle oppressive mechanisms siphoning their aspirations to live.

Without state and federal support, Black gay/SGL males built social support infrastructures ranging from health to employment services. Cohen further explains how Black gay/SGL males wrestled with living amid relegation to a social death: "And yet, out of their vulnerability, they built generative possibilities for community, care, and survival."[9] In *Evidence of Being: The Black Gay Cultural Renaissance and Politics of Violence*, author Darius Bost lists Black gay men's support and advocacy groups, particularly those referenced in his book detailing Black gay/SGL male activism. The earliest organization is the Committee of Black Gay Men, organized in 1979 to combat negative stereotypes of Black gay identity.[10]

CALENDAR

OCTOBER   1991

GMAD is a support group dedicated to consciousness raising, HIV/AIDS prevention education and the development of the Gay and Lesbian community . GMAD is inclusive of **ALL** African, African-American, Carribean, Carribean-American, Latino or Black Gay Men.

GMAD meets weekly, generally Fridays at 8:00 p.m. at sites around N.Y.C. Meetings focus on different topics of relevance to the GMAD community. Special meetings are planned from time to time. Voter registration information and forms; safer sex information, condoms; counseling and referrals are available at all meetings. Feel free to bring snacks and non-alcoholic beverages to share.

Help GMAD serve your needs by sharing your ideas for activities; or by helping to organize specific events. You can make suggestions by mail or phone (see address and numbers below), or volunteer in person at any GMAD meeting.

You are welcome to attend GMAD activities without becoming a member, but we encourage you to support the organization soon. Your donations keep us alive as a vital organization in NYC. Dues are $25/year individual, $12.50/ six (6) months individual, $45/year households and $15/year students/PWAs/persons with fixed incomes. Membership forms are available at most meetings, or you may write to the address below.
Call (212) 226-4433, (212) 226-4462, (718) 788-7124
or write to
80 Varick Street, Suite 3E, New York, NY 10013
For HIV/AIDS information and counseling, please call (212) 239-1796

Figure 13.1.   Gay Men of African Descent Calendar, 1991. Used with permission of Jajuan Johnson.

Other organizations that developed in the 1980s and early 1990s were Gay Men of African Descent in New York, ADODI, Inc. in Philadelphia, Brother to Brother in San Francisco, Black Men's Xchange located in multiple states, NAESM in Atlanta, and the Brotha's and Sista's, Inc. of Little Rock, Arkansas.[11] There is also an expansive list of Black gay men who worked alongside Black lesbians to establish the foundation for health service organizations targeting marginalized communities: Mario Cooper, Reverend Earl Bean, Gil Gerald, Dr. Ron Simmons, Phil Wilson, and A. Billy Jones to name a few. Earlier organizations set precedent for recent advocacy groups such as Chicago Black Gay Men's Caucus, Thrive SS in Atlanta, Impulse Group, the Arkansas Black Gay Men's Forum based in Little Rock, and the National Black Gay Men's Advocacy Group. The existence of these groups affirms that Black gay/SGL males were not passive bystanders amid the uncertainties of an epidemic, but they pressed through fear, denial, and physical limitations to heal their communities.

Conclusion: The Charge

In his last public speech, a eulogy in hindsight, Black gay novelist and professor Melvin Dixon provided a charge to writers, thinkers, and cultural producers at the OutWrite 1992, a conference for gay and lesbian writers: "You, then, are charged by the possibility of your good health, by the broadness of your vision, to remember us."[12] He further added: "Closure involves the will to remember, which gives new life to memory."[13] I assert this is one of the functions of public history. The past represents an ending but also an opportunity to make more inclusive histories. Visitors to museums and cultural centers are not merely seeking answers about the past but are grappling with current phenomena. In the blog piece "Race, Homosexuality and the HIV/AIDS Epidemic," historian Dan Royles amplifies the voices of Black gay/SGL males fighting to end the epidemic: "What, then, is the way forward? Black gay men have been telling us—it's long past time that we listened."[14] As the public seeks to understand more about the past and present realities of HIV/AIDS considering the COVID-19 pandemic, this chapter invites museum curators and interpreters to explore the stories of Black gay/SGL males.[15]

Notes

1. Kevin Mumford, *Not Straight, Not White: Black Gay Men from the March on Washington to the AIDS Crisis* (Chapel Hill: University of North Carolina Press, 2016), 188.
2. Centers for Disease Control and Prevention, Morbidity and Mortality Weekly Report: MMWR, "AIDS: The Early Years and CDC Response," October 7, 2011, https://www.cdc.gov/mmwr/preview/mmwrhtml/su6004a11.htm.
3. Peter Piot, *AIDS: Between Science and Politics*, trans. Laurence Garey (New York: Columbia University Press, 2011), 3.
4. Cathy Cohen, *The Boundaries of Blackness: AIDS and the Breakdown of Black Politics* (Chicago: University of Chicago Press, 1999), 79.
5. "AIDS and the Black Community," CSPAN, March 16, 1998, https://www.c-span.org/video/?102633-1/aids-black-community&fs=e&s=cl.
6. Steve Hendrix, "Robert Rayford Challenged Narrative about the Pandemic," *The Washington Post*, May 15, 2019, https://www.washingtonpost.com/history/2019/05/15/mystery-illness-killed-boy-years-later-doctors-learned-what-it-was-aids/.
7. Centers for Disease Control and Prevention, Morbidity and Mortality Weekly Report: MMWR, "Update: Trends in AIDS Among Men Who Have Sex with Men, United States, 1989–1994," June 2, 1995, https://www.cdc.gov/mmwr/preview/mmwrhtml/00037153.htm. Also see, Centers for Disease Control and Prevention, "Fact Sheet: HIV African American Gay and Bisexual Men," December 2021, https://www.cdc.gov/hiv/pdf/group/msm/cdc-hiv-bmsm.pdf; and Christopher Chauncey Watson et al., "Development of a Black Caucus within the HIV Prevention Trials Network (HPTN): Representing the Perspectives of Black Men Who Have Sex with Men (MSM)," *International Journal of Environmental Research and Public Health* 17 (2020): 871.
8. Essex Hemphill, *Ceremonies: Prose and Poetry* (San Francisco: Cleis Press, 2000), 27.

9. Cathy J. Cohen, "Democratic Futures: Race, Resistance and the Political Vulnerabilities," 2022 Ryerson Lecture, filmed May 19, 2022, University of Chicago, https://www.youtube.com/watch?v=mqd5zDAHcp0.

10. Darius Bost, *Evidence of Being: The Black Gay Cultural Renaissance and the Politics of Violence* (Chicago: University of Chicago Press, 2019).

11. Bost, *Evidence of Being*, 142.

12. Melvin Dixon, "I'll Be Somewhere Listening for My Name, Gay, Lesbian, Bisexual, Transgender: Literature and Culture," *A Special Issue of Callaloo* 23, no. 1 (2000): 81.

13. Dixon, "I'll Be Somewhere," 81.

14. Dan Royles, "Race, Homosexuality and the AIDS Epidemic," *Black Perspectives*, July 16, 2017, https://www.aaihs.org/race-homosexuality-and-the-aids-epidemic/.

15. Readings that serve as useful introductions to Black gay history include E. Patrick Johnson, *Sweet Tea: An Oral History of Black Gay Men in the South* (Chapel Hill: University of North Carolina Press, 2008); and Dan Royles, *To Make the Wounded Whole: The African American Struggle against HIV/AIDS* (Chapel Hill: University of North Carolina Press, 2020). "A Timeline of HIV/AIDS" helps establish context, available at https://www.hiv.gov/hiv-basics/overview/history/hiv-and-aids-timeline.

CHAPTER 14

Becoming a Scientist

Untangling the Roles of Chemistry Sets

Debra A. Reid

DOES YOUR museum have a chemistry set in collections? If so, you might immediately associate it with educational toys of a certain era. You might exhibit it with other items associated with youth culture, or with other product packaging of a particular era. You might be concerned about chemicals packaged with the set, or you might re-create the experiments published in the instructional booklets that the sets usually included. Beyond these enlightening efforts, however, lies new opportunities for inquiry. Digging deeper into chemistry sets, or science sets more generally, makes clearer the ways that race, class, and gender affected public perceptions of science and of scientists historically and today.

A correlation between prepackaged chemistry sets and interest in science existed. James Franck (1882–1964), a Nobel Prize–winning physicist, recalled thinking about physics as a child, but making experiments in chemistry. "There was a chemical set one could get. And I got things of that type." Robert F. Curl Jr. (1933–2022), an American Nobel Prize–winning chemist, remembered that his parents gave him a chemistry set when he was age nine. Curl explained that "Within a week, I had decided to become a chemist and never wavered from that choice."[1]

We may never know the brand names, but Franck's set may have been distributed by Christian Vetter, who marketed science teaching materials prepared by Ludwig Hestermann for use in classrooms and homes. Decades later Curl may have chosen from two popular options. The Porter Chemical Company promoted its Chemcraft Chemistry Outfit as educational, entertaining, and life-changing: "Experimenter today—Scientist tomorrow." Not to be outdone, Chemcraft's major competitor, the A. C. Gilbert Company, linked its Gilbert Chemistry Outfit to "Great fun learning Chemistry of everyday things."[2]

Some sets in museum collections have provenance that link the user's decisions to pursue chemistry, biology, mathematics, or engineering to the sets. The question remains, did the interest precede the gift of the set, or did the set inspire the interest? What other factors affected levels of interest and could lead to a career in science?[3]

Inspiring Future Chemists

The answer to this "chicken-or-egg" question lies in a survey of known inspirations. Assessment of chemistry sets alone tells only part of the story, often one of exclusion rather than access. Additional research, drawing on oral interviews and scientists' autobiographies, confirm that a combination of interest and need also motivated children to pursue chemistry. Some had more leisure to be curious. Others had real problems to solve that could substantially improve their health and well-being (hair- and scalp-care products, for example). Access to chemicals, enthusiasm for experimentation, an inquisitive nature, and access to a library also laid the groundwork for careers in science.

Curiosity may best describe George Washington Carver (1864?–1943) and his fascination with the natural world. He collected plants from a young age, hiding them in a secret garden because "it was considered foolishness . . . to waste time on flowers." He became known as the plant doctor, nonetheless, and parlayed his love of plants into an advanced degree in agricultural science, the first African American to do so. He specialized in plant pathology and mycology. Problem-solving may best describe Annie Malone's motivation. Annie Turnbo Malone (1869?–1957) experimented with compounds to improve scalp health and encourage hair growth for people of color. She and Carver used their chemistry acumen to address challenges that Black Americans faced.[4]

Many scientists acknowledge the importance of libraries, either public or family holdings, as inspirational. George R. Carruthers (1939–2020), Black physicist and engineer, stressed the importance of libraries to his coming of age as a scientist. Roald Hoffman (1937–), a Jewish American Nobel Prize–winning theoretical chemist, born in Zolochiv, Poland (now Ukraine), remembered reading two books, translated into German, in a refugee camp after World War II. One was Ève Curie's *Madame Curie* (1937) and the other was Rackham Holt's *George Washington Carver: An American Biography* (1943).[5]

These examples indicate that inspiration came from many sources. Chemistry outfits, in fact, supported some but not all youthful science experiences. The examples in science and history museums can prompt exploration of the many additional avenues that youth took to engage with science historically and help us be more sensitive to the avenues of engagement available today.

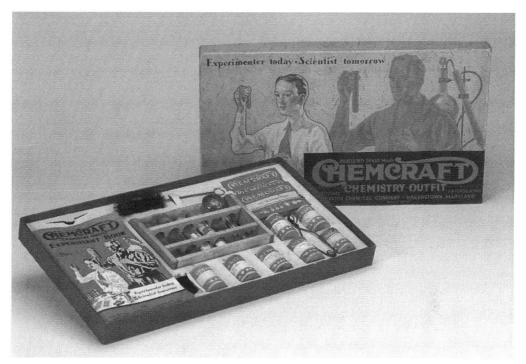

Figure 14.1. Chemcraft Chemistry Outfit, in original box with one bottle brush, one spoon, one set of tongs, one test tube rack, three vials of chemicals, two small bottles, one test tube, seven wood jars of chemicals, three packages of litmus paper, and one experiment book (1934). THF304449. From the Collections of The Henry Ford.

Chemistry Sets as Social and Cultural History

Chemistry sets flourished during the 1920s as educators and social reformers sought to increase access to hands-on educational activities, believing that these could reduce vagrancy and idleness among children. They encouraged teachers to use sets to inform classroom instruction and to support student participation in science clubs and fairs. Parents could direct children keen on experimentation toward a structured experience, with directions provided.[6]

Rebecca Onion puts chemistry sets into the larger framework of popular science in her book, *Innocent Experiments*. She reinforces the importance of self-discovery, citing science writer Steven Silberman, who opined that "The lure of do-it-yourself chemistry has always been the most potent recruiting tool science has to offer." Many who study educational toys survey the history of these sets, indicating the ways that packaged sets became less potent and perhaps less appealing as policy required manufacturers to produce safer kits. Onion cited *New York Times* science writer Malcomb W. Browne, who lamented the decline of the "malodorous and occasionally alarming" home chemistry lab. Chemistry professor Michelle Francl, in her essay "Homemade Chemists," worries that commercial chemistry sets, prepackaged with essentially harmless chemicals, "have deprived a generation of potential chemists of pivotal, if not necessarily essential, experiences." She cites studies that confirm that out-of-school engagement with science sustains interest and motivates students to pursue science.[7]

Minimizing caustic and potentially dangerous chemicals in chemistry sets reduced their inherent risk, but other products, including atomic energy kits, captured the popular imagination. The Science History Institute includes more than 250 science kits and toys. The virtual exhibition *Science at Play* addresses chronological change and topics as wide ranging as "enterprise, aspiration, discovery, magic, and mayhem." It can inspire history museums to seek deeper historical and cultural context for the sets in their collections.[8]

Chemistry sets produced during the 1930s and 1940s can launch an exploration into who had prominence on product packaging, and who did not. Sets conveyed the reality of a white-male-centric profession. The Chemcraft Chemistry Outfit, packaged in 1934, featured a white boy in the foreground, with a white man in the background. A Chemcraft Chemical Magic set from the early 1940s retained messaging, "Experimenter today— Scientist tomorrow," along with the white boy and man juxtaposition, and added "Educational" and "Entertaining." Chemistry sets similar to this were advertised in the 1941 FAO Schwarz catalog at a retail price of $3.50 each. Robert Curl Jr. recounted the influence that a set he received around 1943 had on his career path. More research into users, and into local business history, public and private education systems, and science history, can expand our understanding of who worked in chemistry, and who taught chemistry, during the time frame of kit popularity.[9]

Kits continued to feature white male subjects after World War II. These included Chemcraft atomic energy kits released in 1947, the Atomic Energy Kit No. U-238 issued by A. C. Gilbert Company in 1950, and "professional" chemistry sets marketed through the heyday of kits during the 1950s. Graphic artists employed by at least one company, however, incorporated girls into kit illustrations by 1947. The Practical Chemist, produced by Science Creations, Inc. in 1947, featured a girl and boy on the product packaging (both white), but still stressed gender divisions of labor in chemistry. The girl could apply her understanding to tasks such as making cold cream and window cleaner (as well as inks and ink eradicator). Boys could wow audiences with chemistry magic tricks and could remove stains and polish silver with their chemical concoctions.[10]

What significance should we place on such depictions? Certainly these reflect race, class, and cultural norms of an era, but they cannot be dismissed as benign. Instead, IDEAS (inclusion, diversity, equity, and access to science) begs us to analyze the graphic art depictions and think about how we today fill the voids. This forces us to consider these kits as more than inspiration. They also reinforced the exclusivity of science as an occupation for white Americans. Chemists of color who worked as teachers, college professors, researchers, and inventors confirm that a color line existed. They faced issues that their white counterparts did not, and persevered.[11]

Chemistry sets contrast with proactive efforts today to engage everyone in science. Science organizations document pay rates and other inequities as they recognize women and people of color who serve as role models. Much of this may be front and center during February and March (Black History Month and Women's History Month, respectively), but that is a start toward year-round inclusion, diversity, equity, and access in sciences.[12]

Chemistry sets, and science sets more generally, provide opportunities to inform formal science education. Faculty in high-school and college chemistry classrooms might have their own collection. Opportunities abound to compare and contrast holdings and engage students in the assessment. This might address alchemy and the mysterious nature of chemistry, and the use of chemistry to create synthetic materials. Chemistry sets often included magic in their name and in directions in instruction books, guaranteed to thrill an audience. It could also lead to discussions of materials science. Historical resources might pique a local chemist's curiosity, and that might lead to a pop-up exhibit, lecture, or series of programs directed toward a wide audience. Beyond chemistry sets, other toys, the Barbie scientist doll representing British space scientist Maggie Aderin-Pocock, for instance, affirms commercial response to public expectations about more diverse representations of scientists. This could open other avenues of engagement. All can facilitate transdisciplinary exchange between the Two Cultures, science and history.[13]

Notes

1. This chapter uses "set," "outfit," and "kit" interchangeably, in keeping with the term used by manufacturer or user. Interview of James Franck and Hertha Sponer Franck by Thomas S. Kuhn and Maria Goeppert Mayer on July 9, 1962, Niels Bohr Library & Archives, American Institute of Physics, College Park, MD, https://www.aip.org/history-programs/niels-bohr -library/oral-histories/4609-1. Franck received his Nobel Prize in 1925, as a German scientist. As a Jewish scientist, he left Nazi Germany in 1933, relocating to Johns Hopkins University, then the University of Chicago, where, in mid-1943 he worked on the Manhattan Project as the director of the Chemistry Division of the Metallurgical Laboratory and chaired a committee that issued the *Report of the Committee on Political and Social Problems*. This report "argued against the use of the bomb on an occupied city (instead arguing that it should be used as a demonstration first)," as narrated by Alex Wellerstein, *Restricted Data: The Nuclear Secrecy Blog*, January 11, 2012, "The Uncensored Franck Report: 1945–1946," https://blog.nuclearsecrecy .com/2012/01/11/weekly-document-9-the-uncensored-franck-report-1945-1946/. "Robert F. Curl, Jr., Biographical," The Nobel Prize in Chemistry 1996, https://www.nobelprize.org /prizes/chemistry/1996/curl/biographical/.
2. Portable chemistry set, "Light & Heating," by Christian Vetter, Hamburg, formerly Ludwig Hestermann teaching institute, ca. 1880, Science Museum Group, https://collection .sciencemuseumgroup.org.uk/objects/co8597421. The Chemcraft Chemistry Outfit by the Porter Chemical Company, not dated but including an instruction book copyrighted in 1934, THF304449, The Henry Ford, Dearborn, Michigan. The Gilbert Chemistry Outfit for Boys by the A. C. Gilbert Company, copyright 1943, Science History Institute, Philadelphia, PA, https://digital.sciencehistory.org/works/8c97kr42x. John Tyler, *The Chemcraft Story: The Legacy of Herald Porter* (Haworth, NJ: St. Johanne Press, 2003). "Public Collections of A. C. Gilbert Company Products and Materials." A. C. Gilbert Heritage Society, fifth ed., February 24, 2022.
3. John Brockman, *Curious Minds: How a Child Becomes a Scientist* (New York: Pantheon Books, 2004), features twenty-seven scientists and what inspired them. James Greenhoe (1932–) used this Chemcraft Chemistry Outfit, dated 1937–1946, made by The Porter Chemical Co, Hagerstown, MD. He became an engineer, THF97436, The Henry Ford, Dearborn, Michigan.

The Science History Institute, Philadelphia, PA, has an identical kit: https://digital.science history.org/works/zk51vh70w. Greenhoe also had a toy microscope, made by Ring Optical Company, THF303641, The Henry Ford, Dearborn, MI. Robert L. Veenstra (1943–2015), an electrical engineer, used the Gilbert Chemistry Experiment Lab, dated around 1956, made by A. C. Gilbert Company, TH157626 and THF157627, The Henry Ford, Dearborn, MI. Kenneth Head (1953–) worked as an emissions certification engineer at Chrysler Corporation. He received a Gilbert ChemLab4 as a Christmas gift around 1965, ObjectID 2023.6.1, The Henry Ford, Dearborn, MI. Curator Jeanine Head Miller collected his reminiscence: "I asked for the chemistry set as a Christmas gift. I think using it gave me more of an interest in science as I continued my education through junior high, senior high, and college."

4. Gary R. Kremer, *George Washington Carver: A Biography* (Santa Barbara, CA: Greenwood, 2011), 8; 31–38; Gary R. Kremer, ed., *George Washington Carver: In His Own Words*, second ed. (Columbia: University of Missouri Press, 2017); Mark R. Hersey, *My Work Is That of Conservation: An Environmental Biography of George Washington Carver* (Athens: University of Georgia Press, 2011); Suzanna Maupin Long, "Annie Minerva Turnbo Pope Malone," *Dictionary of Missouri Biography*, Lawrence O. Christensen, William E. Foley, Gary R. Kremer, and Kenneth H. Winn, eds. (Columbia: University of Missouri Press, 1999), https://missouri encyclopedia.org/people/malone-annie-minerva-turnbo-pope.

5. Interview of George Carruthers by David DeVorkin on August 18, 1992, Niels Bohr Library & Archives, American Institute of Physics, College Park, MD, www.aip.org/history -programs/niels-bohr-library/oral-histories/32485. *Roald Hoffman on the Philosophy, Art, and Science of Chemistry*, Jeffrey Kovac and Michael Weisberg, eds. (New York: Oxford University Press, 2012), 1, 9; and Hoffman's Nobel Prize biographical essay, https://www.nobelprize.org /prizes/chemistry/1981/hoffmann/biographical/.

6. C. P. Russell, "The American Boy Wants a Practical Toy," *Printer's Ink* 110 (February 1920): 45–48; Sarah Zielinski, "The Rise and Fall and Rise of the Chemistry Set," *Smithsonian Magazine* blog, October 10, 2012, https://www.smithsonianmag.com/science-nature/the-rise -and-fall-and-rise-of-the-chemistry-set-70359831/; Elizabeth Berry Drago, "Rebel without a Chemistry Set," Science History Institute blog, https://sciencehistory.org/distillations /rebel-without-a-chemistry-set; and Sevan G. Terzian, *Science Education and Citizenship: Fairs, Clubs, and Talent Searches for American Youth, 1918–1958* (New York: Palgrave Macmillan, 2013). Philip Ball, "All Set for Chemistry," blog, Chemistry World (Royal Society of Chemistry, May 14, 2015), https://www.chemistryworld.com/features/all-set-for-chemistry /8536.article.

7. Rebecca Onion, *Innocent Experiments: Childhood and the Culture of Popular Science in the United States* (Chapel Hill: University of North Carolina Press, 2016), 166–69; and Michelle Francl, "Homemade Chemists," *Nature Chemistry* 4, no. 9 (2012): 687–88. Surveys of educational toys usually cover chemistry sets, noting change over time and larger social and cultural history lessons they can help teach. Gary Cross, *Kids' Stuff: Toys and the Changing World of American Childhood* (Cambridge, MA: Harvard University Press, 1997); Hillary Ellis, "Colorful Chemistry Sets," The Strong National Museum of Play, blog, February 16, 2018, https://www .museumofplay.org/blog/colorful-chemistry-sets/; Karen Hewitt and Louise Roomet, *Educational Toys in America: 1800 to the Present* (Burlington: The Robert Hull Fleming Museum, University of Vermont, 1979); and Richard O'Brien, *The Story of American Toys: From Puritans to the Present* (New York: Abbeville Press, 1990).

8. *Science at Play*, Science History Institute, blog, https://www.sciencehistory.org/science-at-play-digital.

9. Salim Al-Gailani, "Magic, Science and Masculinity: Marketing Toy Chemistry Sets," *Studies in History and Philosophy of Science* 40 (2009): 372–81, https://doi.org/10.1016/j.shpsa.2009.10.006. Al-Gailani's work with chemistry sets reached a wide audience through Alex Hudson, "Whatever Happened to Kids' Chemistry Sets," BBC News (August 1, 2012), https://www.bbc.com/news/magazine-19050342. Chemcraft Chemistry Outfit, the Porter Chemical Company, Digital resource ObjectID 91.337.1, The Henry Ford, Dearborn, MI, in original box, around 1934. Chemcraft Chemical Magic, 1937–1946, ObjectID 91.87.61, The Henry Ford, complete in its original red wood case with yellow cardboard lining, blue painted shelves, and a blue painted test tube rack.

10. Porter Atomic Energy Kit (late 1940s, 1950s) and Gilbert U-238 Atomic Energy Lab (1950–1951), ORAU Museum of Radiation and Radioactivity, Oak Ridge Associated Universities, https://www.orau.org/health-physics-museum/collection/toys/porter-atomic-energy-kit.html and https://www.orau.org/health-physics-museum/collection/toys/gilbert-u-238-atomic-energy-lab.html, respectively. Gilbert Chemistry Experiment Lab, circa 1956, ObjectID 2012.49.1, The Henry Ford, advertised as a "professional" grade chemistry set designed to facilitate study of chromatography, atomic energy, glass blowing, hydroponics, secret writing, crime detection, electricity, magic tricks, plastics, and mineralogy. Atomic Energy Lab by A. C. Gilbert Co., 1950, ObjectID 2003.101.3, The Henry Ford. Advertising stressed the ways that Gilbert engineers worked with nuclear physicists to create a "complete Laboratory everyone can [use to] explore the mysterious universe of the Atom—with complete safety! . . . One of the Gilbert instruments alone, the Geiger-Mueller Counter, enables you to prospect for Uranium . . .!" Jane E. Boyd, "Science at Play: A History of Chemistry Sets," blog, Science History Institute, https://artsandculture.google.com/story/science-at-play-science-history-institute/nAXRsuiYboFtKA?hl=en. The Practical Chemist is available digitally via the Science History Institute at https://digital.sciencehistory.org/works/rx913q48q.

11. For more on the theory and practice of accounting for racism, sexism, ageism, xenophobia, and other bigotry in science historically and today, see Karen-Beth G. Scholthof, "IDEAS: Inclusion, Diversity, Equity, and Access to Science," the concluding chapter in this book.

12. "ASM's Focus on Women Microbiologists," American Society for Microbiology, Web page, 2023, https://asm.org/IDEAA/Resources/Women-Microbiologists?utm_medium=email&utm_source=rasa_io&utm_campaign=newsletter.

13. Museums with chemistry sets in collections have opportunities to engage with faculty and students, such as a blog about a visit of students studying "Electromagnetism: Physics, Magic, Religion" during the Fall 2022 semester. Course faculty took students to the Yale University's Peabody Museum to study scientific instruments relevant to course content, including a Gilbert Junior Microscope & Lab Kit, dated 1958. See Alexi Baker, "Teaching with Scientific Instruments at Yale, Fall 2022," blog issued January 23, 2023, https://peabodyhsi.wordpress.com/2023/01/23/teaching-with-scientific-instruments-at-yale-in-fall-2022/. Faculty also require chemistry kits for specific courses. These provide nice comparisons to the earlier iterations marketed as educational toys. The links between magic, alchemy, and chemistry run deep. See Lawrence M. Principe, *The Secrets of Alchemy* (Chicago: University of Chicago Press, 2013); and Jennifer M. Rampling, *The Experimental Fire: Inventing English Alchemy, 1300–1700* (Chicago: University of Chicago Press, 2020). Standards for "new" chemistry sets

indicate an effort to reach more future scientists, as evidenced by Philip Ball, "MEL Chemistry Sets," *Chemistry World* (December 9, 2015), https://www.chemistryworld.com/culture/mel-chemistry-sets/9245.article. Wellcome Group, UK reviews educational kits, i.e., "10 Best Science Kits for 6 Year Olds," reviewed June 22, 2022 (2023 updated), https://wellcometreeoflife.org/best-science-kits-for-6-year-olds/.

CHAPTER 15

The "Gene Gun" and Genetic Engineering

Unpacking the Science

Karen-Beth G. Scholthof and Debra A. Reid

FLIP THE switch on the wall and the light comes on over the kitchen sink. What seems almost magical is actually the results from a three-part system of electrical generation, transmission, and distribution. The system is complex, and while we may not see the beginning or the middle, we know what should happen at the end—light. This chapter addresses a scientific discovery that starts a system that affects all of us (much as electricity factors into our lives). The discovery is genetic modification or genetic engineering of plants. It is today present everywhere at the beginning of our food, fuel, and fiber production systems. Guests to museums and historic sites may have strong and perhaps even polarized opinions about genetic engineering (GE), yet few may grasp the process nor recognize the tools that made GE commercially viable. The following provides an overview of the science, the objects, and the relationships that interpreting the science can illuminate.

First, some history and context for the experimentation that laid a foundation for GE research. During the 1950s, scientists documented the three-dimensional structure of deoxyribonucleic acid (the double helix). This discovery confirmed the function of DNA as a genetic marker and proved that the double-helix structure replicates itself. Documenting the role DNA played in heredity changed the way we understand all living things. Not only did this discovery affect disciplinary practices in anthropology, biology, botany, ecology, and forensic science, among others, but it revolutionized human and veterinary medicine and plant and animal propagation.

Bacteria and their viruses, bacteriophages, were key tools for these mid-twentieth-century explorations. Today genetic engineering is used to explore the inner life of a cell and to determine how DNA mutations could alter an organism's physical features: the phenotype (e.g., the observable and enormous differences in dogs) and the genotype (the passing down of the phenotype to its progeny). The genotype is also part of familiar projects such as the Human Genome Project. This new field of molecular genetics was predicated on decades of cumulative advances resulting from decades of research in biology, chemistry, biochemistry, and physics.

During the 1980s plant biology underwent a scientific revolution predicated on the tools of molecular biology.[1] In addition to the fundamental research underway at universities, scientists at multinational companies with agricultural portfolios made crucial discoveries in plant biotechnology. One goal was to genetically engineer crops so they could fend off the ravages of insects and tolerate certain herbicides used to kill weeds in fields.[2] Seed companies were keen to vertically integrate seed and agrochemical sales to strengthen their commercial portfolios. Meanwhile, the US government actively developed regulations to monitor and control the types of plants and pathogens that could be used in the field.[3]

What tools did scientists use during the 1980s as they experimented with plant genetic engineering? Certainly, their work during the age of molecular biology differed from the work that plant breeders undertook a century before, during the heyday of plant hybridization and the development of hybrid corn. Becoming more familiar with the molecular biology tools of plant science can support comparisons with the more common historic evidence of the age of hybridization. This comparison can inform us about science and reduce barriers to understanding how scientists worked as they articulated hypotheses, conducted their experiments, and oftentimes abandoned discoveries as goals, interest, and technology changed.

Traditionally, agricultural practices and plants had adapted to local conditions. Prior to 1987, horticulturalists and plant breeders, such as Luther Burbank and W. Atlee Burpee, used selective breeding, including cross-pollination of different varieties or species, to produce hybrids. These hybrids were invented to improve appearances or solve problems—thus a seed is a powerful exhibit object to discuss technological and scientific changes. Seed packets, commercial seed catalogs, and bulletins published by technical experts conducting research on experimental farms represent some of the resources available to document this phase of plant breeding.

It's important to note that selective breeding can change plant characteristics, and even plant genetics, but it takes many growing seasons to realize goals. By the early 1980s, GE-based plant breeding was envisioned as the new "green" revolution, perhaps supplanting the tedious decades-long process of developing plants for the garden and field with desired features, such as disease and insect resistance, new flower colors and patterns, vegetables with longer shelf-life, increased nitrogen fixation capabilities, enhanced photosynthesis, and the ability to tolerate drought or salty soils.[4]

What exactly was needed during the 1980s to produce a GE plant? Plants are uniquely different from animals in that any single plant cell can be regenerated into a new plant. Forcing "foreign" DNA into a plant cell and integrating the foreign DNA within the plant's chromosomal DNA modified the plant's genetics and resulted in a regenerated plant (roots, leaves, stem, flowers, and seeds) with that genetic modification intact. But what tools would work best for the process of genetically engineering a plant?

Plant transformation with a bacterium (*Agrobacterium tumefaciens*) was used for many plants, but it was not very efficient for genetic modification of all plants, including small grains (e.g., maize, wheat, millets, rice). A team of research scientists at Cornell University began working on a new tool to facilitate genetic engineering of plants in 1983. Three members of the team filed their first patent in 1984 but continued to refine and improve their invention over the next three years.[5] By May 1987 they were ready to announce the new technology, the gene gun. The team, including molecular biologists, plant breeders, and mechanical engineers, created a modified air rifle that used carbon dioxide (CO_2) cartridges to "shoot" tiny tungsten beads coated with tobacco mosaic virus RNA into onion bulb skin.[6]

Research proved that this method of gene transfer worked, that RNA or DNA could be introduced, and the GE plant cells propagated. This held the potential to rapidly modify plant species, especially rice, wheat, or corn (maize) as well as soybeans and cotton. The technique paved the way for further experimentation with genetic modification of agricultural commodities.

The homegrown Cornell gene gun process was renamed "biolistics" by 1991, a blending of biology and ballistics. Then, E. I. du Pont de Nemours registered "biolistics" as a trademark by 1995 and licensed the "particle gun," now known as a Helios Gene Gun, and the "biolistic particle delivery system" for sale exclusively through Bio-Rad. The California-based company had marketed the first commercially successful "gene pulser" in 1987. No longer was plant-focused genetic engineering undertaken by a cobbled-together device. Instead, as the technology changed, the gene gun became a black box, that is, a part of a system noted for inputs or outputs, but with inner workings little understood.[7]

Today in the United States more than 90 percent of cotton, soybean, and corn (maize) is genetically modified to protect plants from insect pests, plant pathogens, or herbicides. Approximately one dozen commercially available crops are listed as bioengineered; all are

Figure 15.1. Particle gun used in Charles S. Gasser's laboratory, 1990. THF185472. From the Collections of The Henry Ford. Gift of Charles Gasser Laboratory, University of California, Davis.

generally regarded as safe, with no known adverse health effects for humans or animals that consume the products.[8] Since January 2022, in the United States, manufacturers must disclose (label) GE foods and food products.

Genetic modification is one fulcrum around which concerns pivot over food security, the environment, farmlands, and human and animal health. Mostly, GE for human health, such as insulin and other medicines, is accepted by the public. And the recent gene-editing technology CRISPR (clustered regularly interspaced short palindromic repeats) is increasingly being used and accepted to modify human, plant, and animal cell DNA, ushering in a new era of therapeutic genetics and bioethical concerns.[9] For GE, the controversy has focused on crops and concerns about how GE crops may affect human health and increase use of synthetic chemicals because the crop engineering facilitates synthetic-chemical use. Responses to the science includes public resistance in the form of the slow food and locavore movements, as well as the public embrace of climate-smart agriculture intent on reducing greenhouse gas emissions, and often aided and abetted by "smarter" GE crops

Figure 15.2. Graphic organizer to highlight science topics relevant to corn/maize (*Zea mays*). Other organizers could itemize social science topics such as economics and political contexts, or humanities and liberal arts topics such as depictions of corn in art, poetry, literature, glass (whisky bottles shaped like ears of corn), etc. Created by Karen-Beth G. Scholthof. Used with permission.

and robotic technology. These topics will not diminish in importance anytime soon, and museums and historic sites will do well to consider them in the context of their institution's mission and reason for being. Overall, the broad topic of crop-focused genetic engineering, a recent development (since the 1980s), provides opportunities to discuss controversial aspects of modern food, fiber, and fuel production, namely, genetic modification, synthetic chemical applications, and environmental consequences.

How can we use this object to explain a particular science and scientific changes when the basic technology is not familiar to most visitors at a historic site? We suggest that the "black box" aspect can be beneficial as it necessarily generates comments, opens new avenues to discuss biotechnology, and brings forth discussions about American innovations and the quest for discovery. Here, our intent is to encourage curators to locate scientific objects that can be meaningfully discussed within the context of changes in science and technology or everyday living at their historic site.[10]

It is also critical to use historical resources (archival and artifactual) to put changes in science and technology into context. The comparison of pre- and post-molecular-biology plant systems provides a case in point. Consider it an opportunity for discovery. Invite experts representing both perspectives to have a discussion and field questions. Few institutions have the first generation of gene guns or biolistic particle delivery systems. However, many more may have evidence of pre-molecular-biology-age plant engineering. This is a good place to begin the process of discovery and to realize that science is a continuing, unfinished process.[11]

Notes

1. For an overview of plant sciences, see Karen-Beth G. Scholthof's "Plant Sciences: A Brief History," in *A Companion to American Agricultural History*, R. Douglas Hurt, ed. (Hoboken, NJ: Wiley Blackwell, 2022), chapter 13, https://doi.org/10.1002/9781119632214.ch13.
2. Here, the terms genetically engineered (GE), bioengineering, transgenic, and genetically modified are used interchangeably.
3. This followed from earlier decisions to regulate and monitor the uses of genetically modified DNA following the Asilomar Conference on Recombinant DNA. See Paul Berg, "Asilomar 1975: DNA Modification Secured," *Nature* 455 (2008): 290–91. There is a wealth of material on this period, including Peter Pringle, *Food, Inc.: Mendel to Monsanto—Promises and Perils of the Biotech Harvest* (New York: Simon & Schuster, 2005); Belinda Martineau, *First Fruit: The Creation of the Flavr Savr Tomato and the Birth of Biotech Food* (New York: McGraw-Hill, 2001); Pamela C. Ronald and Raoul W. Adamchak, *Tomorrow's Table: Organic Farming, Genetics, and the Future of Food* (New York: Oxford University Press, 2018). The US Food and Drug Administration has tutorials on agricultural biotechnology, including genetic engineering of foods, at https://www.fda.gov/food/consumers/agricultural-biotechnology; and "DNA From the Beginning" at Cold Spring Harbor Laboratory provides animations, videos, and lesson plans on genetics and molecular biology at http://www.dnaftb.org.
4. The Flavr Savr tomato is an interesting case study. It became the first commercially available genetically modified food, though it had a short commercial life, from 1994 through 1997. It was genetically engineered to rot slower after picking, thus extending its shelf life. To

learn more about the twenty-million-dollar research product that produced the Flavr Savr, see Martineau, *First Fruit* (2001).

5. In 1984 the patent was filed with the US Patent and Trademark Office, then modified in 1991 and 1995 as "Method for Transporting Substances into Living Cells and Tissues and Apparatus Therefor," Inventors: John C. Sanford, Geneva; Edward D. Wolf, Ithaca; Nelson K. Allen, Newfield, all of N.Y. Assignee: Cornell Research Foundation, Inc., Ithaca, N.Y. Appl. No.: 670,771. Filed: Nov. 13, 1984. Patent No. 4,945,050, Date of Patent July 31, 1990.

6. T. M. Klein, E. D. Wolf, R. Wu, and J. C. Sanford, "High-Velocity Microprojectiles for Delivering Nucleic Acids into Living Cells," *Nature* 327 (May 1987): 70–73.

7. DuPont, https://patents.google.com/patent/US5036006A/en (accessed October 24, 2022). DuPont commercialized the device for use in the research laboratory. Today, the refined handheld apparatus known as the Helios Gene Gun is exclusively sold through Bio-Rad (Hercules, CA; https://www.bio-rad.com).

8. Salmon is the only bioengineered animal product that falls under the new labeling regulations. See the USDA's Agricultural Marketing Service "List of Bioengineered Foods" at https://www.ams.usda.gov/rules-regulations/be/bioengineered-foods-list. This site also has a "determination tool" that will inform the producer if a food or product must be disclosed as bioengineered; see https://www.ams.usda.gov/rules-regulations/be/zingtree (accessed October 24, 2022).

9. For more on the social and scientific aspects of gene editing, see Walter Isaacson, *The Code Breaker: Jennifer Doudna, Gene Editing, and the Future of the Human Race* (New York: Simon & Schuster, 2021); and Jennifer A. Doudna and Samuel H. Sternberg, *Crack in Creation: Gene Editing and the Unthinkable Power to Control Evolution* (Boston: Houghton Mifflin Harcourt, 2018). Videos and tutorials on CRISPR-Cas9 gene editing and advances in the technology and bioethics of the practice are explored at Innovative Genomics Institute at https://innovativegenomics.org.

10. These "quests for discovery" can, of course, include non-biological everyday objects. For example, a CD player and its "black box" technology, including the development of laser and chip technology, data storage and retrieval, and audio output.

11. The Smithsonian Institution, National Museum of American History collections include Biolistic Particle Gun, Prototype II, ObjectID 1991.0785.02, available at http://americanhistory.si.edu/collections/search/object/nmah_1167050; and Biolistic Particle Gun, Prototype III, ObjectID 1991.0785.01, http://americanhistory.si.edu/collections/search/object/nmah_1167048. The Henry Ford collections include the particle gun used in Charles S. Gasser's laboratory at the University of California-Davis, 1990, ObjectID 2020.37.1, https://www.thehenryford.org/collections-and-research/digital-collections/artifact/502082; and DIY Gene Gun (aka Particle Delivery System) with Accessories, 2019, ObjectID 2022.78.1, https://www.thehenryford.org/collections-and-research/digital-collections/artifact/525308, made by The Henry Ford's historic operating machinery staff based on directions provided by Jay Hanson, Kyle Taylor, and Arnie Wernick, "How to Build and Use a Gene Gun," O'Reilly (May 2, 2016), https://www.oreilly.com/content/how-to-build-and-use-a-gene-gun/. For an earlier "how to" example, see Amit Gal-On, Eti Meiri, Chassia Elman, Dennis J. Gray, and Victor Gaba, "Simple Hand-held Devices for the Efficient Infection of Plants with Viral-encoding Constructs by Particle Bombardment," *Journal of Virological Methods* 64 (1997): 103–10, https://doi.org/10.1016/S0166-0934(96)02146-5.

CHAPTER 16

Know Your Analyst, Know Your Food?

Benjamin R. Cohen

TRANSITIONS ARE hard. That's true personally; it's true environmentally; and it is especially true for political change, where the seedbed for foment and transformation will often lie underground, invisible to most for years and decades before the clarity of the change pops. This chapter is about a broad transformation in the world of food and farming that indeed took decades. The transition in question occurred across a half century between the end half of the nineteenth century and first half of the twentieth. It concerned how people understood what their food was, if it was safe, and if it was healthy.

There's an enduring question underlying this change: How do we ever know what our food is? When I ask people that, they often look puzzled. "It's common sense," they say, or "I cook a lot so I know," or "It says right on the label." We have a variety of answers at our modern disposal. For most of history, people relied on their senses to answer that question, they trusted the farmer they got the food from (if it wasn't self-provided), and they leaned on their community experience. Late-nineteenth-century industrial patterns disrupted those methods. The other side of the change in this chapter's driving question—how do you know what your food is?—had new scientific work redefining the answer by asking consumers to trust the analyst, not themselves. How did that change happen?

The Disruption of Sensory Confidence

The farming community had been the arbiter of purity for centuries. There's something commonsensical about that statement when you keep in mind that the basic structure of most societies for millennia had been agrarian. People didn't always use the term "pure," which came into greater public consciousness in the English-speaking world in the later

1800s. At that time, there was an identifiable social and public health movement in the United States called "the pure food crusades." In the centuries before that, genuine, true, or natural would have been more common. Regardless of the term, people have consistently worried about food purity.

Imagining the staples of a general store, of milk, bread, grains, butter, and the like, provides readers today with access to the scene. Butter in particular is a good example of how the grower was the arbiter and trusted agent of purity along with agrarian community experience. In this case, cookbooks offer insight on how people understood the truth of their food. Amelia Simmons, the author of the nation's first notable cookbook, *American Cookery* (1796), wrote her recipes and knew her food from "the close familiarity . . . a woman who cooked could have with animal's lives." Knowing the cow's age, disposition, and daily habits influenced how Simmons understood the character and quality of butter from "the consistency, flavor, and color of its milk."[1] Having intimate knowledge of terrain, weather, and neighbors structured familiarity with food quality. Advising readers on pure "sweet butter," for instance, Simmons suggested sending "stone pots to honest, neat, and trusty dairy people." Those honest farmers would bring it back "in the night, or cool rainy morning, covered with a clean cloth wet in cold water, and partake of no heat from the horse, and set the pots in the coldest part of your cellar," to ensure its unadulterated condition.[2] The sense of pure, in other words, derived from the familiar interactions of local life with trusted growers. And when there was deception, because there was also deception in the preindustrial world, you knew where to go for the accusation.

In the craft-based societies before the industrial age people also came to know their food through common sensory measures, what scientists would now call organoleptic metrics. Sight, touch, taste, and smell served as profound measures of food identity.

The value of sensory experience played out in the battle between purity and its enemy, adulteration. Adulteration was a term that lived in the same space as corruption, deception, and contamination. Purity and adulteration were understood through the lens of sensory experiences that didn't yet involve complicated tools or instruments. Those would come later. Those would come when industrial techniques and urban contexts confused people too much about what their food was. They didn't know the grower anymore.

But first, the literature on anti-adulteration would often recommend avoiding adulteration by diligent sensory tests. A guide from 1832 advised readers to choose butter "based on taste and smell," to ascertain fraud in "the quality of flour [by] colour and feel," and to detect adulterated oil by "smell" and "the dullness of the colour." Cookbooks for centuries called on cooks to gauge quality, identity, and safety of food through sensory interactions. "Olfactory vigilance," to quote the French historian Madeleine Ferriere's lovely phrase, paired with the decisiveness of sight and taste in securing the confidence of the customer.[3]

The color palette of foods served as a complex means for such security. Yellow could be valuable or cautionary, for example. White was ideal for the common household ingredient lard, with butchers frequently using candles and lighting to showcase the color and hide any tint of yellow. But yellow was preferable for butter or other dishes that could benefit from the addition of saffron, or "parsley juice for green" or "sunflower for purple."[4] This is to say that appearance helped inform identity. Sellers knew this, butchers knew it, bakers and spice dealers and dairymen knew it.

The hitch was that as the nineteenth century progressed, foods started to come from odd places, far places, confusing places. More people were living in cities and more of them were shopping for food at a grocers' market (we call them grocery stores now), where it was less clear where the food came from. It was becoming less obvious who the grower was or if your senses might be deceived by the ever-increasing chain from what we now call farm to fork. What if you couldn't trust your senses, or your experience, or your grocer?

As it happens, people are clever. They devise methods for understanding things. In the case of food identity, chemists and public health figures developed instruments people could use to vet their food's quality and purity, instruments that augmented primary senses.

These methods first came from agrarian communities themselves. This is again more of a commonplace comment than it might seem—as I recalled above, the basic structure of most societies for millennia had been agrarian. When Congress passed legislation for a US Department of Agriculture in 1862, Lincoln would call it "The People's Department." The science grew from that culture.

The sphere of domestic science is a good example. Across the mid-1800s, Catherine Beecher's *Treatise on Domestic Economy* (1841) was the standard-bearer for instructing homes on managing food identity and preparation. By the later 1800s, Ellen Richards's work did well to show the evolution of domestic science from domestic economy, and by the twentieth century, home economics.

Richards was a Boston-based chemist and a self-proclaimed scientist of "oekology"—the "household of nature," as she translated it, before twentieth-century scientists called it ecology. She wrote fifteen books and coedited several more. She was the leader of the new American Home Economics Association and founded the *Journal of Home Economics*. Her *Food Materials and Their Adulterations* hit shelves in 1886 and went through two more editions in the coming decades. It helped that she had grown up in rural Massachusetts working in her father's general store, a countryside analogue to the urban grocers' market.

Producers processed foods in large-scale facilities by the end of the nineteenth century to a degree never before seen. Many readers might associate that phenomenon with a mid-twentieth-century timeline, but the scale, resources, packaging, branding, and processing steps that generally define an industrial process over a craft or artisanal one were all in place at the turn of the century. What we find later is an extension and acceleration of that system, not its beginnings. The point is, that transition period across those early industrial decades confused eaters. People like Richards helped them gain confidence in their food. Their lack of confidence had come from the loss of stable trust mechanisms where farming communities were the arbiters of purity and identity.

There is a much fuller story to tell about efforts to build a domestic science as a solution to the problem of food identity and trust.[5] The relevant part here is that chemists, home economists, and housewives—who were often the same people—were all involved in extensive debate in the later 1800s with grocers, boards of health, and various governing bodies about growing confusion over arbitrating the identity and purity of food. What had been squarely in the purview of the household was becoming a matter for larger structures of governance and technical analysis.

THE "PURE FOOD" SITUATION AT A GLANCE. Fig. 170.

Figure 16.1. Clipping from a late 1890s newspaper showing USDA scientist Harvey Wiley framed as the proprietor of A. Devil & Co., with adulterants under the counter meant by the artist to convey the problems of trust in grocers' markets. Courtesy of the Wiley Papers, Accession #MSS45690, Library of Congress, Washington, D.C.

A New Century

A host of cultural, agricultural, technological, and political developments by the early twentieth century shaped changes in how the everyday household, cooker, and eater could trust their food. Dynamic cultural changes put stress on existing modes of interaction between people and, more to our point, stressed the confidence to know who was growing and selling food. Dramatic demographic changes in the United States wrought by expansion to the west, along with urbanization into cities and from the countryside in the east and the intake of vast immigrant populations, furthermore, upended the patterns of well-heeled community norms. Agricultural changes added their part. For one thing, settlement and colonial expansion ("Manifest Destiny") had put new lands under greater cultivation across the broader Midwest and Plains states in ways that would only continue in the twentieth century. Global commodity trade routes also increased the availability of foreign ingredients, crops, and processed foods. Technological systems like rail distribution, mechanized farm equipment, refrigeration, shipping, and factory production were inspired by the cultural and agricultural changes and, in turn, pushed them forward.

A useful point in the historical timeline to see the changing times came in the early 1900s, with two notable events in 1906. One was the Pure Food and Drug Act, which Congress passed and President Roosevelt signed that summer after decades of prior national legislative failures (189, to be exact). The other had more literary allure—Roosevelt read it right away—which was Upton Sinclair's *The Jungle*. Sinclair published it that winter and saw it very quickly become a bestseller. *The Jungle*'s imagery is still so vibrant that many works about the era of pure and adulterated food will pin its resolution to Sinclair's journalism.

Those were bigger, bolder, more visible examples. It's the less visible part of scientific infrastructure that factors more into our story, though. In particular, the rise of an analytical industry across the latter decades of the nineteenth century made it possible for twentieth-century governance to draw from, rely upon, and then center the authority of chemical analysis in questions about food identity. The way to know if a food was what its sellers claimed was to analyze it with an instrument. The results were often advertised as part of a label on a product. That presumes there were labels, there were products, there were analysts, and there was a reason to want to know these things. All of those took work, giving substance to the aforementioned seedbed of foment and transformation that rode along less visibly underground.

Recentering food governance in urban stores was key. The broad arc of that transition in food provisioning—where do you get your food—moved from rural communities, where the self-reliant farm family or general store provided the main access point for food, to urban centers. That shift had already begun in European nations throughout the nineteenth century; the same pattern accelerated in the United States as the twentieth century began.

Ellen Richards and a cohort of writers and domestic economists were tackling the problems of food knowledge for the household, as noted above. They sought to bring chemical detection to the kitchen itself. Richards, for example, helped open the New England Kitchen in 1890, famous for its dedication to nutrition, working-class diet, and respite for the poor. This kind of work also trickled into homes through popular magazines like *Godey's* (founded in 1830), *Ladies' Home Journal* (1883), and *Good Housekeeping* (1885).

Along the way, various governing entities began to tackle the same problem. Outside the home, boards of health worked on developing standards and practices for confirming the safety of foods. Occasionally, grocers hired chemists to vet the identity of their foods in an effort to garner the trust of customers. This happened too with a new industry of local analysts, who set up shop much as a pharmacist or accountant might, plying their trade for a market that demanded it. It happened at new corporations as well. Procter & Gamble led the way with in-house labs, soon followed by meatpackers like Armour & Co. and producers like Heinz.

In the mid-1800s, there was no stable infrastructure of chemical shops, trade papers, labs, boards of health, or grocers' shops. When Congress passed its Pure Food and Drug Act in 1906, there was. They could rely on an active infrastructure of chemical and scientific detection. Consumers would also find packaged foods, brands, trademarks, logos, labels, and analyses where there had been little of those before. Changes in answering the question "how do you know what your food is?" were profound, having moved the seat of power from the farming community to the analyst's lab.

Conclusion

The consumer had begun to understand the veracity of their food with labels and product packages, where certified analyses from invisible but newly trusted agents secured confidence in the food's identity.

By returning to where I started, I'm abbreviating a thicker and more tangled story to provide two sides of a transition. It was one whose front side came from a world where the dominant way to understand food purity was based in the home. It led to a world where food identity and purity were the purview of scientists in a lab. Truth be told, the shift from the grower and eater as arbiter of purity to the scientist and lab was tucked inside a host of larger changes. My main point is that knowing your analyst became as important as knowing your food. A new intermediary had been born. A new way to trust the food you ate came from new forms of analysis and detection that before had been unnecessary, unappealing, or too scattered. Scientific analysis in the form of ingredient labels, nutrient claims, advertised qualities, and regulated marketplaces all formed the foundation for what would only grow in depth and complexity in the century to follow, to this day. The reaction this century to concerns over an industrialized food system—often labeled a local food movement in the first decade of our century—tacitly sought to regain the personal knowledge of food that had grown out of reach. Knowing what your food is in the future could benefit from re-establishing measures of personal knowledge, but at a political scale they will always include some kind of assurance from some kind of extra-household analysis.

Notes

1. Ann Vileisis, *Kitchen Literacy: How We Lost Knowledge of Where Food Comes from and Why We Need to Get It Back* (Washington, DC: Island Press, 2008), 31.
2. Amelia Simmons, *American Cookery, or the art of dressing viands, fish, poultry, and vegetables, and the best modes of making pastes, puffs, pies, tarts, puddings, custards, and preserves, and all kinds of cakes, from the imperial plum to plain cake: Adapted to this country, and all grades of life* (Hartford, CT: Hudson and Goodwin, 1796), 9.
3. *An Enemy to Fraud and Villainy, Deadly adulteration and slow poisoning unmasked; or, Disease and death in the pot and the bottle* (London: Sherwood, Gilbert, and Piper, 1832), 15 and 138. Also see Madeleine Ferrières, *Sacred Cow, Mad Cow: A History of Food Fears*, trans. Jody Gladding (New York: Columbia University Press, 2005). For further detail about the changing concepts of taste and flavor in the early twentieth century, see Nadia Berenstein, "Flavor Added: The Sciences of Flavor and the Industrialization of Taste in America" (PhD diss., University of Pennsylvania, 2017), available at https://repository.upenn.edu/edissertations/2715; and Ai Hisano, "'Eye Appeal Is Buy Appeal': Business Creates the Color of Foods, 1870–1970" (PhD diss., University of Delaware, 2016), available at https://udspace.udel.edu/handle/19716/21122.
4. Ferrières, *Sacred Cow, Mad Cow*, 71.
5. I tell that story in *Pure Adulteration: Cheating on Nature in the Age of Manufactured Food* (Chicago: University of Chicago Press, 2019), from which these examples are drawn. Even that work draws from a host of excellent works on the history of domestic science, nutrition,

and household chemistry. See, for example, Charlotte Biltekoff, *Eating Right in America: The Cultural Politics of Food and Health* (Durham, NC: Duke University Press, 2013); Kristin Hoganson, *Consumers' Imperium: The Global Production of American Domesticity, 1865–1920* (Chapel Hill: University of North Carolina Press, 2007); and Jessica Mudry, *Measured Meals: Nutrition in America* (Albany: State University of New York Press, 2009) are good sources for more depth on the topic.

CHAPTER 17

Science and Progress in the Kitchen

Forks, Eggbeaters, and Sporks

Karen-Beth G. Scholthof and Debra A. Reid

SCIENCE IS oftentimes narrated as a march of progress, ranging from the stepwise evolution of man from ape; from fire to nuclear fission; from animal pelts to synthetic fibers. This Whiggish version of advancement displaces previously hard-won knowledge, skills, and historical tools made by humans over eons, in favor of a straightforward, or linear, notion of progress. This false dichotomy of old as antiquated and obsolete, and new as improved and efficient, is worthy of exploration as we consider how to interpret science at museums and historic sites. A methodology exists to support such exploration—the social construction of technology. It consists of a framework that helps us avoid presentism when analyzing artifacts. It does so by stressing the context of the artifact's time (history) and place (environment), the social and cultural milieu that affected inventors, adopters, users, and modifiers, and the role of science across the technology's life span.

As an interpretative case study, we present what may seem a frivolous example, but it is one that shows the tension of defining "advances" in science and technology: the eggbeater. A stick, a pronged stick, the fork, then the whisk, provide the entry-point technology: age-old tools sufficient for whisking or beating eggs or mixing liquids. Then, the eggbeater. In addition to whisking, an eggbeater was promoted as a labor-saving culinary tool for the housewife that, for instance, quickly produced whipped cream or whipped egg whites. The first US patents for eggbeaters were granted in 1856. By 1859 the Monroe eggbeater was patented, followed by hundreds of patents for "improved" devices, although the design was only slightly altered. This was the standard tool until the mid-twentieth century, when

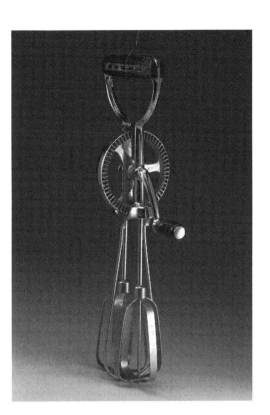

Figure 17.1. Eggbeaters were advertised and used as time-saving tools for nearly a century before the arrival of the electric eggbeater and mixer for home use. Examples of these objects (left to right to bottom) include a patented eggbeater, an Ekco stainless steel eggbeater (1945–1955), and a "Magic Maid" electric mixer, ca. 1935 with bowls, juicer and maceration attachments. Patent 23,694, J. F. and E. P. Monroe, April 19, 1859, U.S. Patent Office; THF319904, from the Collections of The Henry Ford; and THF134132, from the Collections of The Henry Ford.

commercially produced electric mixers were marketed to the home cook, although they looked remarkably similar to Monroe's device.[1]

Within the framework of interpreting science, what can we learn from the evolution of the eggbeater? When we itemize changes in eggbeater technology, we may conclude that the electric eggbeater is the most advanced and that it has evolved from the hand-powered versions. This approach to interpreting an artifact type as a stepwise progression from past to present, with the most recent labeled the most advanced iteration, is derisively named "Whiggish history" or "presentism" by historians of science. Such simplification leads us to devalue the nuance and complexity of the science that influenced the inventor, adopter, and adapter. And science influenced all steps in the innovation process, including patenting, refining, processing, distributing, adopting, and adapting. Saying "no" to presentism—resisting the urge to simplify change as progress—means that we can value artifacts as products of their time and can better chart the discoveries that led to early and subsequent iterations.

How do we place the form and function within the context of the time period without defaulting to our present point of view? As argued by Angélica Vasconcelos and colleagues, an interdisciplinary approach is important in avoiding Whiggish history and the perils of presentism—they urge us to be mindful of the importance of "embracing the expertise of specialists from different relevant fields."[2] To embrace the past requires research—Who are our actors? Why are they using the tools? Who is developing and patenting improvements? Are there socioeconomic or cultural aspects related to tool adoption? Who benefits from adoption of new, or modified tools? Who is the end user: A commercial kitchen, a restaurant chef, a collector, a home cook, or an enslaved laborer?[3]

Perhaps we should begin with the fork, then address the whisk and eggbeater. The utilitarian kitchen fork began with two tines and use in food preparation rather than food consumption. This changed as cultures developed foods that they could not consume easily with the more common knife or spoon. Bee Wilson credits pasta makers with fork modifications. Science relevant to this change involves the selective breeding of wheat to secure a pasta flour and the chemical reactions of flour and water (and perhaps an egg) to yield a dough that, when further processed, yielded a durable noodle that lent itself to consumption with a three-tined fork. The material science of metals and finishing materials add additional opportunities to discuss science around the kitchen tables with the fork as the focus.[4]

A derivative of the fork is the spork. The hybridization of the spoon bowl and the fork tines was promoted as performing both features in a single utensil. It too has a long history. In 1874, a combined knife, fork, spoon utensil was patented in the United States. And a slew of modifications emerged. Yet a spork does not work better than its antecedent utensils: the spoon and the fork.[5] Curiously, the spork remains in use around the globe. This begs the question as to why this technological "advancement" remains in circulation. Arguments include reducing plastic waste in the fast-food restaurant industry, ensuring incarcerated persons do not have access to weapons (forks or knives), and reducing backpack weight when hiking or trekking. If the spork is a niche object, then it can be argued, contextually and historically, that it is not an advancement in culinary utensils.[6] It is as a "modern" present-day object that is not necessarily "building" on the past nor reflecting an "evolution" of a technology. Thus, the spork, based on our use of the tool, does not represent inevitable progress—it is a tool that has become embedded in our material culture without commentary. Of course, this is a simplified example, but it is intended to challenge us to look more carefully at advertised "new and improved" advancements in science and technology in our everyday world and to critique and understand how to move away from the "easy" history of this scientific march of progress.

Now, let's return to the eggbeater. What can we say about science (and its technology) when viewing an eggbeater? First, is this an improvement in technology? Where was it manufactured? What was the "evolution" of the components—the gear mechanism, especially. What is the physics and chemistry behind the eggbeater—that is, how does it work to beat egg whites into a meringue, or cream into whipped cream? And is it a better tool than a whisk? For example, to whisk, one hand is used, leaving the other to hold the bowl. The eggbeater requires two hands—what anchors the bowl? A whisk "feels" more controlled compared to the eggbeater that flings eggs to the sides and over the bowl and requires a flat bottom to work efficiently as a tool. A whisk also can be used to stir; an eggbeater cannot.

As narrated by Bee Wilson, there was an eggbeater bubble in the decades bracketing the twentieth century. It was the tool for the modern housewife or those in her employ. It was to be a time-saving tool, yet anyone taking a turn at an eggbeater knows it is an enormously frustrating device to use. So, here again, we find that some material objects are popularized even though they are not a better tool than their antecedents. Not until the commercialization of the electric eggbeater or mixer did we see some of these issues resolved.

As the United States became electrified in urban and rural areas, with the rise of the refrigerator, and other technological changes in the home, electric appliances became popular for their potential to save time, if not labor, required to prepare three meals a day. The electric mixer had attachments devoted to specific purposes from beating and whisking, to grinding and extruding, and kneading. They may have resulted in more cleaning and a need for more countertop space, but these items do seem an advance in technology. These tools, some or all of which would be found in a house museum, or their replicas, can be used to test or demonstrate how well they function at specific tasks.

From this, the point is to "do" a bit of experimentation, a fundamental science methodology. Practice artifact assessment by observing, analyzing, and discovering the nuances of tools and related technology of a past time, and in the context of the science of cooking. Then repeat the experiment with related tools and technologies of another time, and in the context of the science of cooking. What breakthroughs exist? What social and monetary costs did these breakthroughs entail? Ruth Schwartz Cowan argues that food preparation changed significantly with the microwave, but the technology did not reduce labor. Instead, it facilitated a shift in domestic labor away from cooking and toward work outside the home (among other things). But as long as foods facilitate cultural cohesion, science facilitates food procurement, processing, and preservation, and science supports the materials deployed as tools that facilitate those tasks.[7] Are you ready to whisk these details into an interesting interpretation of science and domestic life?

Notes

1. Ruth Schwartz Cowan applied the social construction of technology model of analysis to refute notions of labor-saving devices in *More Work for Mother: The Ironies of Household Technology from the Open Hearth to the Microwave* (New York: Basic Books, 1983); and Ruth Schwartz Cowan and Matthew H. Hersch, *A Social History of American Technology*, second ed. (New York: Oxford University Press, 2018). These works provide a model for how to think differently about technology that can inform how to think differently about science and its influence on domestic life and decorative arts.

   For a fascinating introduction to history and science of foods and their preparation, see Harold McGee's *On Food and Cooking: The Science and Lore of the Kitchen* (New York: Scribner, 2004). The history of culinary and household utensils is the subject of Henry Petroski's *The Evolution of Useful Things* (New York: Knopf, 1992), especially chapter 1, "How the Fork Got Its Tines;" and Bee Wilson's *Consider the Fork: A History of How We Cook and Eat* (New York: Basic Books, 2012). Importantly, Petroski asks why Western tradition uses the fork and Chinese tradition uses chopsticks. This serves to stress, for us, the need for interdisciplinary thinking when approaching an object as "simple" as a fork.

2. Angélica Vasconcelos, Alan Sangster, and Lúcia Lima Rodrigues, "Avoiding Whig Interpretations in Historical Research: An Illustrative Case Study," *Accounting, Auditing & Accountability Journal* 35 (2022): 1402–30; quote 1403. For more on reevaluation of experiments from past to present, see Karen-Beth G. Scholthof et al., "Practicing Virology: Making and Knowing a Mid-Twentieth Century Experiment with Tobacco Mosaic Virus," *History and Philosophy of the Life Sciences* 44 (2022): 3 and citations therein.

3. Perhaps the most famous American kitchen is that of chef and cookbook author Julia Child. At the Smithsonian, Child's kitchen is used to contextualize American culinary history. The kitchen is on permanent exhibit at the National Museum of American History, Washington, DC, https://americanhistory.si.edu/food/julia-childs-kitchen.

4. Petroski distinguishes between ceremonial forks and food forks in *The Evolution of Useful Things*, chapter 1; and Wilson reports that eating pasta in Italy was a turning point, with the development of three-tine forks (*Consider the Fork*, 192).

5. Other spork-like utensils included an ice-cream scooper and terrapin fork (to dine on freshwater turtle soup or gelatinized mock-turtle soup from calves' heads). The utility of the spork can be experimentally tested. Obtain a spoon, fork, and spork and evaluate their usefulness in the kitchen: beating an egg, twirling spaghetti, slurping soup, scooping ice cream.

6. For further analyses of Whiggish history and presentism, see: Oscar Moro Abadía, "Beyond the Whig History Interpretation of History: Lessons on 'Presentism' from Hélène Metzger," *Studies in the History and Philosophy of Science* 39 (2008): 194–201; David Alvargonzález, "Is the History of Science Essentially Whiggish?" *History of Science* 51 (2013): 85–99; Stephen G. Brush, "Scientists as Historians," *Osiris* 10 (1995): 214–31; Jane Maienschein, Manfred Laubichler, and Andrea Loettgers, "How Can History of Science Matter to Scientists?" *Isis* 99 (2008): 341–49; Ernst Mayr, "When Is Historiography Whiggish?" *Journal of the History of Ideas* 51 (1990): 301–9; and Trevor J. Pinch and Wiebe E. Bijker, "The Social Construction of Facts and Artefacts: Or How the Sociology of Science and the Sociology of Technology Might Benefit Each Other," *Social Studies of Science* 14 (1984): 399–441.

7. Cowan, *More Work for Mother*.

SCIENCE AND HISTORY MUSEUM EDUCATION

In Part IV, the contributors share case studies that feature immersive activities, accessible artifacts, and public engagement efforts that prompt deep thought and dynamic action that extends beyond surface engagement with science interpretation. Bethann Garramon Merkle explores the interplay between art and science and the blurred lines between memory and interpretation. She reveals how new connections can be had by thinking about how the interpretation of science has much to do with building relationships and sharing knowledge between visitors and interpreters and beyond to broader communities. For Debra Reid, relationships between interpretations of science and history museum education can be valued literally as much as intellectually. She traces the dynamic nature of school gardens as artifacts of science and the multisensory connections that visitors can make by exploring this history as well as how museums and historic sites can benefit by participating in experiential learning activities such as "edible education."

Another key way to offer dynamic interpretive experiences for visitors is to find artifacts that had a similar purpose in the past and use them to help make new connections in the present. David Vail's chapter on the USDA *Yearbooks of Agriculture* highlights how an easily overlooked set of government publications can actually reveal much about scientific theories, practices, debates, and technologies that came to define the previous two centuries. Vail shows that while the *Yearbook of Agriculture* had an explicit goal of an accessible agricultural science for producers throughout the country, each volume often covered a variety of scientific specialties, experiments, and technologies published for the general public. For Brian "Fox" Ellis, scientists themselves have important stories to tell. Ellis explores the research, the artifacts, and the interpretive mindset and vocabulary necessary to embody a particular scientist from the past for audiences. Ellis traces the complex interpretive efforts required to bring these historical characters to life in engaging, entertaining, and educational ways. Robert Oleary and Matt Anderson highlight the multifaceted role of science in the permanent exhibit *Driven to Win: Racing in America* at The Henry Ford Museum of American Innovation. Both authors explain the interpretive process in presenting the science and technology of racing, including how race cars themselves can serve as material culture artifacts for these kinds of exhibits. Debra Reid concludes this section with a chapter on nature study that opens access to the interpretation of science for museums and historic sites, through an objective of citizen science or experiential learning. These chapters bring us full circle to convergence education and using multidisciplinary approaches to merge the Two Cultures—putting some STEAM into STEM.

CHAPTER 18

Integrating Art and Science to Effectively Share Knowledge

Bethann Garramon Merkle

PICTURE THIS. You're in an education room in a museum. The room is well-appointed, with plenty of lighting, comfortable seating, and handy work surfaces. The room is also clearly well-stocked with materials of all sorts. But, it does not have any windows. And the facilitator says, "Sketch a tree." That is your only instruction. Someone asks, "What kind of tree?" Another, "How big?" The facilitator only responds, "That's up to you. Just sketch a tree. Don't overthink it." There are no trees to look at, so you scribble down a tree from memory. It looks like a triangle, a stereotypical conifer. You notice that your neighbor has drawn something that could be a fruit tree, or a poofy cloud on a post. You both agree: You have no idea how to draw. Hopefully the facilitator doesn't expect much!

And then the facilitator says, "Okay! Now, sketch a specific tree that was special to you as a child or is special now." Big gulp. You have exactly the tree in mind. Perhaps you grew up near a forest. Perhaps you grew up in an urban environment, and there was just one, iconic tree in your neighborhood. Maybe you recently planted a tree to commemorate someone's birth or death. Whatever the case, you can see this tree vividly in your mind's eye. But getting it down on paper?!?

You can't just duck out, so you labor over what the trunk looks like, how the branches are shaped, what kind of leaves the tree has. And far too soon—you haven't come close to drawing the whole tree—the facilitator asks everyone to pause. You are then prompted to talk with your neighbor about the tree you sketched. Most of your conversation is about the memories you each have of these trees, why you care about the tree, and if you've seen it recently. And, you apologize to each other: for the amateur nature of your drawings, for omitting all sorts of details, especially the ones you can't quite remember. How the branches connect to the trunk. The actual shape of the canopy of the tree. What sorts of scars or marks were on the trunk. Whether any other animals (besides you) seemed to value the tree or use it for food or shelter.

You're still not quite sure what the facilitator is up to, but you enjoy thinking about your special tree for a few minutes. You might have preferred to just reminisce or write notes about it, though. Drawing is hard, and you wouldn't want anyone to see your sketch. It looks like a little kid drew it! Why did you need to draw anyway?

The facilitator seems to anticipate your questions. "Now, it's time to write some notes. Add some context and details to your sketch. Use words to clarify things that didn't come through in your sketch." And, your memories hit the paper in a flood—the color of the leaves or needles at different times of year. The sound of the wind going through in the winter. How you used to interact with the tree. Where parts of the tree "should" be but don't actually look that way in your sketch.

"And now, write down at least one question you have." The facilitator continues, "What's missing? What can't you remember? What would you have to go back to that tree and look at closely, in order to depict it accurately?" And your list swells: no idea what the buds looked like. Did it flower? How big was it, really, compared to memory? Is it still there?

The facilitator asks you to discuss what was similar and what was different about what you made for "draw a tree" versus "draw a specific, special tree." There's a hush, and then everyone starts calling out things like "cloud tree," "It's a cartoon," and "The specific tree is way more detailed."

"Yes, exactly," says the facilitator. And they remind you of something obvious that feels somehow profound: A blank page is blank. And it clicks—your expectations influence both your memories and your new experiences or efforts to learn something new. Your expectations are perhaps more influential than the reality you actually encounter. That's why looking for memory or accuracy or fact on a blank page can be frustrating. And yet, this frustration is avoidable if you are attentive and observant, if you are open to noticing and learning from your surroundings and experiences. It is possible to try something new without "failing" if you moderate your expectations and acknowledge you have arrived with preconceptions.

When the facilitator next prompts you to go out into the museum, with its interactive exhibits and indoor and outdoor displays, you're intrigued. You're actually willing, now, to try to sketch things that catch your attention. And, you've been primed—by sketching your tree from memory—to double-check your perceptions of the museum exhibits. You can use your sketches and notes to keep you tuned in to what's really there, not just what you anticipate you'll see.

As this example[1] shows, what museum and historical site visitors experience is contingent. Our interpretation, memory, conclusions, and even learning hinge on our prior knowledge, social pressures, and even convenience.[2] In reality, human nature is at odds with many of the goals of museums and historical facilities and the ways of knowing that are conveyed through such institutions. These are places, materials, and ideas that are intended to simultaneously entertain and educate. And yet, most of us are unaware of how our biases, expectations, and social positioning influence how we interact[3] with settings like museums.

This paradox is especially acute in settings where science is a central theme or subject. Most science, in the United States at least, is conveyed in a manner rooted in three assumptions: (1) science is objective,[4] (2) this objectivity is not just desirable, it is paramount,[5] and (3) science is a universal social good. And yet, history contradicts these assumptions with deep and mounting evidence that Euro-colonial science has been exploitative, damaging, and exclusionary across cultures, regions, socio-economic circumstances, and time frames.[6]

Only a few examples are needed to emphasize the lack of objectivity in, and exclusionary nature of, science. Indeed, science history and current practices are far too often characterized by exploitation, appropriation, and suppression. Examples include explicit situations such as when Johns Hopkins researchers exploited Henrietta Lacks, a Black woman who died from a devastating, aggressive cancer for which she sought treatment at Johns Hopkins. Researchers there harvested some of the cancer cells from her without permission.[7] They propagated the cells into the HeLa cell line now worth billions of dollars; a resource, which numerous courts have ruled is *not* the property of her family, but of the biomedical companies who commodified her illness. Other examples are more nuanced and explicitly intersect with the arts. For example: Maria Sibylla Merian appropriated knowledge from her slaves in Suriname to "discover"[8] and elegantly illustrate insect metamorphosis and the concept of ecology. And, the Vatican suppressed Galileo's illustrations and dissemination of his realization that the Earth actually revolves around the Sun by threatening him with torture and death (ultimately scaled back to lifelong house arrest). These examples make it clear: both art and science were historically dominated by Eurocentric attitudes that were frequently dismissive, exclusionary, and worse.

As we dig deeper into the whitewashing of science and art, it's a fairly grim and intense history; one that may not be appropriate or authorized for all museums and historical settings. Further, we face a conundrum if we focus on the major flaws of science and modern calls for accountability. That is: museums, historical sites, scientists, and advocates for evidence-based decision-making are struggling to compete with short attention spans, alternative "facts," and massive amounts of information hitting people every day.[9]

In such circumstances, leveraging the provocative and contemplative power of arts practices can allow for more nuanced engagement with the positives and negatives of modern science and science history, *if* we hold space for diverse ways of knowing (e.g., Indigenous science). The arts can invite visitors to productively reckon with the benefits and necessities of a paradigm shift. Indeed, today we are seeing scientists attempt more candid, responsive, inclusive, and socially engaged approaches to conducting science.[10] Arts practices can help us consider and value approaches such as co-production, collaboration, and consultation.[11] Arts integration can even form the basis of efforts to integrate multiple ways of knowing into bodies of evidence and research methods that are mutually beneficial.

One such approach is to use drawing as a relationship-building and knowledge-sharing tool. Dr. Jean Polfus and collaborators resorted to drawing in an attempt to find common ground with Dene elders and hunters in Canada's Northwest Territories.[12] Ultimately, drawing and storytelling facilitated the development of Indigenous-driven research methods that respected local taboos, centered Indigenous knowledge, and made possible the collection of important genetic data about caribou biodiversity.[13] Given the diverse histories and affinities of people who engage with museums and historical sites, similar reciprocal processes can be valuable at these locations.[14]

For a museum-based example of arts-informed knowledge exchange, we can consider the ArT STaRTs (Artful Thinking Science Teacher Resource Trainings) program that I codeveloped at the University of Wyoming in collaboration with the UW Art Museum and the UW Science Initiative's Learning Actively Mentoring Program (focused on enhancing science teaching in higher education). As previously noted, an increasingly polarized media and political environment can negatively impact public perceptions of science[15] and bring increased scrutiny to how people fund and do science. Through ArT STaRTs, we are investigating the utility of public scrutiny of art, particularly public art, to serve as a metaphor and lens for science educators and scientists working to make science more accessible and inclusive.

Art-science integration (aka STEAM) is widely touted as a powerful means of enhancing how people engage with and learn science.[16] Similarly, art-science integration is often seen as a compelling way to enhance scientists' creativity. And indeed, "creativity is crucial to the capacity to do science well, to communicate it in compelling ways, and to enhance learning."[17]

It is counterproductive, however, to use "told, not taught" approaches when integrating arts-based practices in classrooms.[18] As previously discussed, the way we teach science can result in people feeling like they do not belong in science, or that science doesn't care about them. Arts experiences can go similarly poorly without deliberate facilitation. Thus, ArT STaRTs workshops aim to build science educators' awareness of both the opportunities and potential challenges (access, sense of belonging and self-agency, etc.) facing learners in art, science, and integrated environments.

At the same time, we are studying the outcomes of these trainings, because there is a documented need for meaningful assessment of art-science integration training efforts,

Figure 18.1.   ArT STaRTs program at University of Wyoming. In the author's codeveloped ArT STaRTs program, STEM faculty discuss connections between artworks and the science subjects they teach. Here, an evolutionary geneticist describes how a mixed-media painting relates to her undergraduate evolution course. Photograph by Bethann Garramon Merkle © 2023.

Figure 18.2. Chemistry of resin sculpting transforms arts educators' understanding of genetic phylogenies, replication, and biodiversity. An arts educator participates in a summer Art-Science Institute co-developed by the author. Here, she learns about chemical processes necessary to create resin sculptures which are part of a project to construct three-dimensional phylogenetic trees of cichlid fishes. Photograph by Bethann Garramon Merkle © 2023.

particularly beyond K–12 classrooms.[19] Such integration is a developing field of research, and it can be a powerful means of implementing the "Broader Impacts" expectations that federal funders increasingly require.[20] Ideally, museums and historic sites using art-science integration approaches will also contribute to efforts to understand and assess the efficacy of such work. Possibilities for doing so include (1) developing robust assessment and evaluation programs internally and then sharing results externally and (2) consulting or partnering with researchers investigating these questions.

The synergies of art-science integration efforts are actually not surprising, though surprise seems to be a common reaction. In reality, art-making and scientific investigation share a kinship[21] of creativity, curiosity, close observation, and questioning. These aspects of inquiry are coupled, in art and science, with description and interpretation that are essential to the advancement of technology, industry, the economy, and indeed, society's self-concept. Furthermore, creativity can be both practiced and enhanced to strengthen science professionals' capacity in academic, industry, and civic spheres.[22] Indeed, agreed-upon definitions of creativity identify it as one's ability to generate ideas that are both novel and usable—a capacity that we must aim for in educational and civic settings. STEAM initiatives aiming to enhance these capacities are increasingly common and thus can provide us with wide-ranging, robust models[23] for implementation and assessment. When melded together, sociocultural, scientific, and arts considerations can result in transformative practice in both art and science.[24] Settings such as museums and historical sites are ideal for drawing people into complex, fully integrated art-science experiences that authentically model the complementary and additive nature of the arts and sciences.

Notes

1. This facilitation sequence is a central part of the drawing trainings and educator coaching conducted regularly by the author. See Bethann G. Merkle, Brian R. Barber, and Matthew D. Carling, "Drawn to Natural History: Enhancing Field Courses with Drawing and Field Journal Instruction," *Natural Sciences Education* 49 (2020): e200019 for a detailed discussion of this approach.

2. For example, see Jonas T. Kaplan, Sarah I. Gimbel, and Sam Harris, "Neural Correlates of Maintaining One's Political Beliefs in the Face of Counterevidence," *Scientific Reports* 6 (2016): 39589.

3. See George Lakoff and Mark Johnson, *Metaphors We Live By*, second ed. (Chicago: University of Chicago Press, 2003); and Daniel Kahneman, *Thinking, Fast and Slow* (New York: Farrar, Straus and Giroux, 2013) for an extended discussion.

4. Adrian Treves, "Scientific Ethics and the Illusion of Naïve Objectivity," *Frontiers in Ecology and the Environment* 17 (2019): 363.

5. Dean K. Simonton, *Creativity in Science: Chance, Logic, Genius, and Zeitgeist* (Cambridge: Cambridge University Press, 2004).

6. For example, Bethann Garramon Merkle, Evelyn Valdez-Ward, Priya Shukla, and Skylar R. Bayer, "Sharing Science through Shared Values, Goals, and Stories: An Evidence-Based Approach to Making Science Matter," *Human-Wildlife Interactions* 15 (2022): 598–614; Liisa Husu, "Gate-Keeping, Gender Equality and Scientific Excellence," in *Gender and Excellence in the Making* (Luxembourg: European Commission Publications Office, 2004): 69–76, https://op.europa.eu/s/xD90; Samniqueka J. Halsey, Lynette R. Stickland, Maya Scott-Richardson, Tolulope Perrin-Stowe, and Lynnicia Massenburg, "Elevate, Don't Assimilate, to Revolutionize the Experience of Scientists Who Are Black, Indigenous and People of Colour," *Nature Ecology and Evolution* 4 (2020): 1291–93; and Rocío Deanna et al., "Community Voices: The Importance of Diverse Networks in Academic Mentoring," *Nature Communications* 13 (2022): 1681.

7. Rebecca Skloot, *The Immortal Life of Henrietta Lacks* (New York: Crown, Penguin Random House, 2010).

8. Elizabeth Polcha, "Breeding Insects and Reproducing White Supremacy in Maria Sibylla Merian's Ecology of Dispossession," *Lady Science* (June 20, 2019), https://www.ladyscience.com/breeding-insects-and-reproducing-white-supremacy/no57.

9. Merkle et al., "Sharing Science."

10. For example, see Katherine Canfield and Sunshine Menezes, *The State of Inclusive Science Communication: A Landscape Study* (Kingston: Metcalf Institute, University of Rhode Island, 2020); Julie Risien and Martin Storksdieck, "Unveiling Impact Identities: A Path for

Connecting Science and Society," *Integrative and Comparative Biology* 58 (2018): 58–66; and Jean L. Polfus, Deborah Simmons, Michael Neyelle, Walter Bayha, Frederick Andrew, Leon Andrew, Bethann G. Merkle, Keren Rice, and Micheline Manseau, "Creative Convergence: Exploring Bicultural Diversity through Art," *Ecology and Society* 22 (2017): 4.

11. Polfus et al., "Creative Convergence."

12. Bethann G. Merkle, "Drawn to Caribou," *American Scientist* 104 (2016): 16–18.

13. Polfus et al., "Creative Convergence."

14. For example, see Katie Christensen, Bethann G. Merkle, and Brenna Marsicek, "Integrating Art, Science, and Community Engagement: The University of Wyoming Art Museum's Ballengée Project," *Informal Learning Review* 148, no. 1 (2018): 21–27.

15. Jay D. Hmielowski, Lauren Feldman, Teresa A. Myers, Anthony Leiserowitz, and Edward Maibach, "An Attack on Science? Media Use, Trust in Scientists, and Perception of Global Warming," *Public Understanding of Science* 23 (2014): 866–83; Matthew C. Nisbet et al., "Knowledge, Reservations, or Promise? A Media Effects Model for Public Perceptions of Science and Technology," *Communication Research* 29, no. 5 (2002): 584–608.

16. National Academies of Sciences, Engineering, and Medicine, *The Integration of the Humanities and Arts with Sciences, Engineering, and Medicine in Higher Education: Branches from the Same Tree* (Washington, DC: National Academies Press, 2018).

17. Stephanie R. Januchowski-Hartley, Natalie Sopinka, Bethann G. Merkle, Christina Lux, Anna Zivian, Patrick Goff, and Samantha Oester, "Poetry as a Creative Practice to Enhance Engagement and Learning in Conservation Science," *BioScience* 68 (2018): 905.

18. Bethann G. Merkle, "Drawn to Science," *Nature* 562 (2018): S8–S9.

19. National Academies of Sciences, Engineering, and Medicine, *The Integration of the Humanities and Art with Sciences.*

20. National Science Foundation, "Perspectives on Broader Impacts," NSF 15-008, 2014, https://nsf-gov-resources.nsf.gov/2022-09/Broader_Impacts_0.pdf.

21. For example, see Januchowski-Hartley et al., "Poetry as a Creative Practice," 905–11; Merkle, "Drawn to Science;" Christensen, Merkle, and Marsicek, "Integrating Art;" and J. H. van't Hoff, *Imagination in Science*, trans. and intro. G. F. Springer (Amsterdam: Springer, 1879; New York: Springer-Verlag, 1967), 1–18.

22. Januchowski-Hartley et al., "Poetry as a Creative Practice."

23. For example, see Niki A. Rust, Lucia Rehackova, Francis Naab, Amber Abrams, Courtney Hughes, Bethann Garramon Merkle, Beth Clark, and Sophie Tindale, "What Does the UK Public Want Farmland to Look Like?" *Land Use Policy* 106 (2021): 105445; Christensen, Merkle, and Marsice, "Integrating Art"; Amy Duma and Lynne Silverstein, "A View into a Decade of Arts Integration," *Journal for Learning through the Arts* 10, no. 1 (2014): 1–18; and Aaron M. Ellison et al., "Art/Science Collaborations: New Explorations of Ecological Systems, Values, and their Feedbacks," *Bulletin of the Ecological Society of America* 99 (2018): 180–91.

24. For example, Polfus et al., "Creative Convergence;" Christensen et al., "Integrating Art, Science, and Community Engagement;" Januchowski-Hartley et al., "Poetry as a Creative Practice;" and Rust et al., "What Does the UK Public Want."

CHAPTER 19

School Gardens and Edible Education

Debra A. Reid

MANY HISTORIC houses, open-air museums, and living history farms maintain kitchen gardens, orchards, and ornamental gardens. These horticultural features often include period-appropriate varieties of vegetables, flowers, and cultivars and are arranged and maintained in keeping with institutional mission. Conversations about "heritage" varieties may emphasize the flavors and textures of the vegetable or fruit in contrast to modern iterations or may stress the human role in maintaining seeds between growing seasons. Science, however, gets short shrift in these interpretive vignettes. Heritage apple trees provide a case in point. You can grow an apple tree from seed, but you replicate apple varieties through cloning, which involves grafting buds and scions onto a hardy rootstock. The clone replicates the parent tree, and this ensures that the fruit from the clone is identical in size, shape, color, and flavor to the parent. Pomological societies promoted apple traits. To meet demand, commercial nurseries hired experts to clone fruit trees and created colorful marketing materials to encourage commercial growers, farmers, and backyard gardeners to invest. Finally, consumers scrutinized the harvested fruit at markets, buying what they needed for baking, frying, pureeing, or enjoying as a dessert apple.

Commodities like apples and other fruit trees can introduce us to the science of plant propagation. Vegetables and flowers help segue to the science that launched and sustained commercial seed production. Adding these concepts to the interpretation of horticultural features can balance the oft-repeated and romanticized interpretations of heritage varieties. This approach can provide critical context to understand horticulture specifically, and agriculture more generally, as an unnatural act that started with domestication millennia ago. The Smithsonian Institution's curator of agriculture, John Schlebecker, explained it this way: "farmers intentionally produce commodities by controlling biological activity as best they can . . . they grow things on purpose." Gardens lay the foundation to discuss agriculture as purposeful management of biological processes.[1]

While gardens already exist in many historic houses and historic sites, school gardens offer additional opportunities to interpret biology and crop cultivation. School gardens, lesser known but significant among educators and reformers for nearly two hundred years, can inspire interpretation of existing kitchen and ornamental gardens and can justify creation of new gardens that add nuance to other types of sites, especially historic schools.

School Gardens

School gardens, along with nature study, emerged as key elements of US educational reform during the nineteenth century. Those who supported school gardens, and nature study more generally, did so because of concerns about retaining agricultural knowledge in the face of industrialization and urbanization.[2]

Advocates also stressed the value of active learning, what today we might call kinesthetic learning. They considered this essential for retaining knowledge about natural and physical sciences, including biology and soil and plant sciences. Advocates also encouraged teachers to incorporate garden-based lessons into art, literature, and business instruction. Interest in school gardens ebbed and flowed in keeping with changing educational theories and methods, often inspired by responses to economic crisis, war, industrialization and loss of farmland, and moral reform.

Local governments established school gardens in Schleswig-Holstein (Germany) by 1814. Similarly, the Swedish parliament called for teachers to create gardens near their residences and engage students in cultivation in 1842. Students learned about gardening as a result (or added to existing knowledge gained through work on the family farms). Teachers benefited from the school gardens as well, securing foodstuffs and flowers and supplementing their income with produce sales, both of which might have made teaching a more viable occupation.[3]

Postwar recovery from the Austro-Prussian War of 1866 and the Franco-Prussian War of 1870–1871, prompted Erasmus Schwab, an Austrian administrator, to advocate for public-school gardens. His pamphlet, *Der volksschulgarten* (1870), spelled out a model for engaging youth in constructing or rebuilding spaces destroyed during the war. His idea spread with the help of Mrs. Horace Mann (Mary Tyler Peabody Mann), who translated a fourth edition of Schwab's *Der volksschulgarten* as *The Public-School Garden*, published in the United States in 1879.

European educators, pressured by industrialization, adopted school garden methodology before educators in the United States realized the potential. Why? Perhaps this reflected the long-term commitment in the United States to agrarianism, the political philosophy that stressed landownership as foundational to citizenship. Most families in the United States, theoretically, had access to traditional garden knowledge (that intangible cultural heritage based in localized practice). Agrarians might argue, why was education in gardens necessary? Urban reformers, on the other hand, warmed to the notion of outdoor education, especially as health benefits became apparent.[4]

In the United States, immigrants spread the school garden idea as educators, administrators, scientists, and medical professionals advocated for open-air education. Teachers experienced in school garden pedagogy, including immigrants who implemented the idea in other countries, may have transferred their theory and method from their homeland to the United States. How this may have affected rural or urban schools serving ethnic populations remains unexplored. That said, agricultural science became more widely known in the United States after the 1860s because of national investment in higher education. The Morrill Land-Grant College Act of 1862 increased opportunities for students to learn agricultural and mechanical subjects and convey that knowledge to family farms and communities of farm families. The Hatch Act of 1887 increased support for applied science and dissemination of findings and recommendations through special bulletins. Scholars acknowledge the mixed influence of this work on farm practices, but the national investment served science well.[5]

Land grant colleges formed at the same time as teachers' colleges, and this facilitated competition between institutions of higher education. The consequence of this for school gardens has yet to be understood, but enthusiasm for nature study in general, and for school gardens, specifically, increased rapidly during the late nineteenth century. Science factored prominently in the curriculum.

Teachers promoted gardening as an avenue toward understanding natural environments. Wilbur S. Jackman, a Harvard College graduate, formulated nature study while teaching high school in California, Pennsylvania. His *Nature Study for the Common Schools* (1891) laid the groundwork for teacher training in a science-based curriculum that immersed students beyond the textbook. Progressive faculty at teachers' colleges (the Cook County Normal School, which became the University of Chicago), private institutions of higher learning, and land grant colleges such as Cornell University in New York State expanded on school gardens as part of nature study. Faculty at the University of Chicago, including Wilbur Jackman, who became dean of the School of Education there in 1904, and educational innovator John Dewey, joined forces. They taught teachers, featured case studies in serial publications, and consolidated those publications into how-to books.[6]

Medical professionals advocated outdoor education for health reasons, especially as scientists confirmed the cause of infectious diseases, including tuberculosis. German-born Sigard Adolphus Knopf studied in California, New York, and Paris, France, earning two MD degrees, and worked in tuberculosis wards in Paris hospitals before returning to the United States to advocate for health reform. He called for rooftop school gardens and outdoor schools to improve student health in his recommendations to counter tuberculosis disease. Some urban reformers partnered with philanthropists to emphasize the benefits of school gardens as a "foe of the 'White Plague'" (tuberculosis) because they guaranteed students "healthy exercise in the open air."[7]

These progressive educators allied with agricultural scientists to reach rural and urban teachers, students, families, and others concerned about youth education in school and beyond. Cornell University, New York State's land grant college, issued nature-study leaflets starting in 1896. Liberty Hyde Bailey, a horticulturalist and dean of the New York State College of Agriculture at Cornell, linked nature study to school gardens in his presentation, "The Nature-Study Movement," delivered to the National Educational Association in 1903.

While the ideal laboratory was the out-of-doors, he claimed that "the most workable living laboratory of any dimensions is the school garden."[8]

Black Education and School Gardens

Concomitantly, George Washington Carver, the first Black American to earn an advanced degree in agricultural sciences, applied nature study while enrolled at Iowa Agricultural College and Model Farm (now Iowa State University), the land grant in Iowa. He wrote about the subject as the agricultural specialist at Tuskegee Normal and Industrial Institute in Tuskegee, Alabama. His pamphlet, *Progressive Nature Study* (1897), laid out a methodology that Carver pursued throughout his career. He stressed nature study and included botanical analysis in his "how-to" instructions. He suggested that teachers have students study flowers and leaves from oak and mulberry trees. He designed the section on "experimental work," however, "to strengthen actual scientific and practical research, and . . . originality, freedom of thought and action." Specimens suggested for analysis included cotton and collard leaves along with other Alabama plants, namely, wild cactus and "arrow-head" or water lily.[9]

Carver's nature study and school garden instructions incorporated lessons in mathematics and business as well as science, composition, and art. One example focused on trees as commercial products. Others stressed soil regeneration through applications of composted leaves and alluvial sediment. These instructions reached readers within rural Alabama and beyond, be they teachers in rural schools or farm families operating as citizen-scientists.

The school gardens became the setting for a quiet revolution because Carver's recommendations could help impoverished students in rural Alabama feed themselves during times of crisis, namely, during the Jim Crow era. All students, whether children of tenants or sharecroppers or landowners, could implement this advice on school gardens, thus overcoming the limitations imposed by many landlords to restrict farm family activities to cotton cultivation. Teachers in rural schools or in institutes such as Hampton (Virginia) and Tuskegee used school gardens to teach students (young women and men alike) about economic development, self-help, and community building.

Carver expanded his "how-to" advice in 1904 with *Nature Study and Children's Gardens*, the second teacher's leaflet published by Tuskegee Normal and Industrial Institute. This pamphlet stressed the ways that gardening at an early age helped instill good habits in children that they could apply into adulthood as members of successful farm families. The third iteration, published in 1910 as *Nature Study and Gardening for Rural Schools*, repeated ideas introduced in the previous two pamphlets. Namely, "a large part of a child's education must be gotten outside of the four walls designated as class room," that "practical Nature study method cannot fail to both entertain and instruct," that nature study "leads up to a clear understanding of the fundamental principles which surround every branch of business in which we may engage," and that "it also stimulates thought investigation, and encourages originality." Through it all, Carver linked lessons in agricultural science to regenerative practices suited to the natural environment and the social and cultural systems that Black families built despite white supremacist restrictions.[10]

Urban Reformers

Urban reformers also adopted school gardens. An exhibition at the Louisiana Purchase Exposition in St. Louis (1904) stressed the benefits of science education through the immersive environments of school gardens. The pamphlet, *School Gardens* (1910), summarized the history and described the ways that the independent Fairview Garden School Association in Yonkers, New York, created a garden school (not a school garden) for use by neighborhood children.[11]

The intensity of school garden formation prompted publishers to release books to meet the growing need. Many repeated advice about how to turn a small amount of land into a lucrative market garden, goals and tactics implemented at Black institutions of higher education across the South. This was a key component of Alpheus Hyatt Verrill's *Harper's Book for Young Gardeners* (1914). Verrill also appealed directly to youth, rather than teachers or parents. He explained this in his introduction: "With modern methods, intensive cultivation, and improved varieties of plants a very small plot can be made to yield a profit. It is to show how much profit and pleasure may be obtained from simple gardens and how to plan, arrange, and care for them properly that this book has been prepared."[12]

Figure 19.1. Medical professionals believed that school gardens could reduce the potential for these children and their teachers to contract tuberculosis. Photograph taken outside Public School 65, Brooklyn, New York, 1890–1915, by Jenny Young Chandler. THF38044. From the Collections of The Henry Ford. Gift of Betty R. K. Pierce.

Verrill's book met a growing need as national and state investments in agricultural extension education increased, and as 4-H clubs reinforced youth engagement in community gardening. Verrill addressed this in his introduction too. "In many places boys and girls have already formed garden clubs and associations, and the results obtained by some of these youthful gardeners would be a credit to professional horticulturists. Scores of first awards and prizes at county and state fairs have been given to boys and girls for exhibits of their garden (i.e., potatoes and tomatoes) and field products (i.e., corn). In many instances crops and gardens have furnished a substantial income for their young owners."[13]

Many remained committed to the supportive role that school gardens played in nature study, geography, aesthetics, and creativity. "The school-garden is an outgrowth of regular school-work; it is one striking phase of the effort to get out of doors, away from books and into contact with the real world. It is a healthy realism putting more vigor and intensity into school-work." That is how Illinois educator Charles A. McMurry explained the concept in *The New Student's Reference Work* in 1909. He also stressed the ways that school gardens could help build moral character and improve country life. As McMurry opined, "The cultivation of plants requires constant attention, forethought, intelligence, self-reliance and a kind of originality; difficulties are to be met and overcome . . . and the child must be intelligent and thoughtful in meeting such difficulties."[14]

School gardens had inspired US educators for at least eighty years by the time Henry Ford incorporated gardens into instruction at The Edison Institute school and Ford Motor Company Trade School, both of which used Greenfield Village in Dearborn, Michigan, as their campus. Clara Ford also advocated for gardening as central to elementary education, and she pursued this as president of the Woman's National Farm and Garden Association. She created plans for a roadside market stand, displayed a scale replica during the 1929 WNFGA conference in New York City, and established the market stand near Greenfield Village during the 1930s as an outlet for vegetables grown by the elementary-school students. The garden stand proved lucrative. The per-student share for sales in 1936 amounted to $25.15.[15]

The model established by school gardens remains vibrant today. Edible Education, a term coined by activist chef and educator Alice Waters, calls for schools to involve students with food in every phase of its production—planting, harvest, preparing, eating, and repeating. Waters launched the idea as part of the Edible Schoolyard Project at the Martin Luther King Jr. Middle School in Berkeley, California, in 1995. There, she and a small group of teachers, parents, and administrators dug up a parking lot and implemented an interdisciplinary gardening-based curriculum that engaged students through hands-on and minds-on instruction.[16]

Waters acknowledges how her years teaching in Montessori schools and working as a chef and restauranteur influenced her vision of edible education. She also recognizes that educators long advocated for school gardens, often linking nature study and practical agriculture in their calls for engaging students in agricultural practices outdoors. Waters's work propelled a revival in school gardens.

The same rationale that reformers articulated a hundred years ago remains vital today. Biology teachers, according to Christopher Riggs and Danielle Lee, recognize the benefits

Figure 19.2.   Greenfield Village School Market (left) and Dearborn Pantry Shelf (right), at the corner of Southfield Highway and Village Road, Dearborn, Michigan, 1934. THF117976. From the Collections of The Henry Ford. Gift of Ford Motor Company.

of garden-based learning. Students retain information that boosts their confidence and translates into increased awareness of their surroundings, not just their assignments. Gardening seems to further social and emotional learning, according to Abby Lohr and other environmental educators, because the process helps students make decisions, build relationships, and manage their time and energy. KidsGardening recognized heritage months with a virtual lecture series, "Culturally Inclusive Teaching in the Garden." The six topics addressed included principles of culturally responsive garden education and Native American, Hawaiian, and Alaskan Native; African American and Black; Hispanic and Latinx; Middle Eastern and North African; and Asian and American Asian perspectives on gardening.[17]

This overview of school gardens confirms historic precedent for incorporating science education into museum and historic site interpretation. The sources document gardens as an essential component of formal outdoor education and of science education within the context of human need. The publications and minute books of state and municipal boards of education and of teacher organizations indicate the pace of adoption, the reasons for opposition, and the social, cultural, economic, and regional distinctions that defined school gardens. Local research can deepen our understanding of how this played out over time. How did places affect the lessons that school gardens taught as reformers, administrators, and teachers aligned natural sciences (i.e., biology, botany, and zoology), medical science, and social sciences to accomplish their reform goals? Today, ecoliteracy amounts to a new articulation of the long history of school gardens as a learning laboratory to teach about food, culture, health, and the environment.[18]

Museums and historic sites can turn local context for school gardens into programs that engage communities. It can support partnerships with groups advocating for edible education and environmental education. It can spotlight community gardens, urban agriculture, and other environmental-awareness initiatives. The multi-event and interdisciplinary seasonal programming that results will enhance students' sense of place and can help them relate science to history and art, and vice versa.

Notes

1. John T. Schlebecker, *The Past in Action: Living Historical Farms* (Washington, DC: Smithsonian Institution, 1967), 1.

2. For an introduction to nature study, see Reid's later chapter in this volume and the "nature study" entry in selected readings. See also *Cornell Nature-Study Leaflets Being a selection, with revision, from the teachers' leaflets, home nature-study lessons, junior naturalist monthlies and other publications from the College of Agriculture, Cornell University, Ithaca, N.Y., 1896–1904*, Nature Study Bulletin No. 1 (Albany, NY: J. B. Lyon Company, Printers, 1904), available at https://www.gutenberg.org/files/43200/43200-h/43200-h.htm. These leaflets became the basis for Anna Botsford Comstock's *Handbook of Nature Study*, first ed. (1911; rev. ed., 1939) and reprinted with a foreword by Verne N. Rockcastle (Ithaca, NY: Cornell University Press, 1986). Cornell University maintains a digital archive that addresses nature study and rural education, available at https://rmc.library.cornell.edu/bailey/naturestudy/naturestudy_2.html.

3. Educators summarized school garden history. James Ralph Jewell offered one of the earliest and most complete overviews in his chapter, "School Gardens," *Agricultural Education including Nature Study and School Gardens*, Bulletin No. 2 (1907), Whole Number 368, Bureau of Education, Department of the Interior, second ed., rev. (Washington, DC: Government Printing Office, 1908), 23–46. Also see Petter Åkerblom, "Footprints of School Gardens in Sweden," *Garden History* 32, no. 2 (Winter 2004): 229–47; and Mary Forrest and Valerie Ingram, "School Gardens in Ireland, 1901–24," *Garden History* 31, no. 1 (Spring 2003): 80–94.

4. Erasmus Schwab, *Der volksschulgarten. Ein beitrag zur lösung der aufgabe unserer volkserziehung*, second ed. (Wien: E. Hölzel, 1873), translated by Mary Tyler Peabody Mann as *The School Garden: Being a Practical Contribution to the Subject of Education* (New York: M. L. Holbrook & Co., 1879).

5. Alan I. Marcus, ed., *Science as Service: Establishing and Reformulating American Land-Grant Universities, 1865–1930* (Tuscaloosa: University of Alabama Press, 2015); and Alan I. Marcus, ed., *Service as Mandate: How American Land-Grant Universities Shaped the Modern World, 1920–2015* (Tuscaloosa: University of Alabama Press, 2015).

6. Sally Gregory Kohlstedt, "Nature, Not Books: Scientists and the Origins of the Nature-Study Movement in the 1890s," *Isis* 96, no. 3 (2005): 324–52; and Sally Gregory Kohlstedt, "A Better Crop of Boys and Girls: The School Gardening Movement, 1890–1920," *History of Education Quarterly* 48, no. 1 (2008): 58–93.

7. Daniela Blei, "When Tuberculosis Struck the World, Schools Went Outside," *Smithsonian Magazine*, Education during Coronavirus: A Special Report (September 1, 2020); Ruth Clifford Engs, "Knopf, Sigard Adolphus," *The Progressive Era's Health Reform Movement: A*

Historical Dictionary (Westport, CT: Praeger, 2003), 195–97; S. Adolphus Knopf, *Tuberculosis as a Disease of the Masses and How to Combat It*, seventh American ed. enlarged and rev. (Chicago: University of Chicago, 1911), 69; and Mrs. A. L. Livermore [Henrietta Jackson Wells], *School Gardens: Report of the Fairview Garden School Association, Yonkers, N.Y.* (New York: Department of Child Hygiene. Russell Sage Foundation, 1910), quote 14.

8. L. H. Bailey, "The Nature-Study Movement," in *Cornell Nature-Study Leaflets*, Nature Study Bulletin No. 1 (Albany, NY: J. B. Lyon Company, Printers, 1904), quote 28.

9. G. W. Carver, *Progressive Nature Study* (Tuskegee, AL: Tuskegee Normal and Industrial Institute, 1897), quote 7.

10. Geo. W. Carver, *Nature Study and Gardening for Rural Schools*, Bulletin No. 18 (Tuskegee, AL: Tuskegee Normal and Industrial Institute, 1910), quote 3.

11. Livermore, *School Gardens* (1910).

12. A. Hyatt Verrill, *Harper's Book for Young Gardeners: How to Make the Best Use of a Little Land* (New York: Harper & Brothers, 1914), quote xvii.

13. Verrill, *Harper's Book for Young Gardeners*, quote xviii.

14. C. A. McMurry, "The School-Garden," in *The New Student's Reference Work for Teachers, Students, and Families*, Chandler B. Beach, ed., Frank Morton McMurry, assoc. ed., Vol. 4 (Chicago: F. E. Compton and Co., 1909), 1690–91, quotes 1690 and 1691, respectively.

15. Jim McCabe, "Clara Ford's Roadside Market: A Small Building with Big Aspirations," The Henry Ford, Dearborn, MI, September 9, 2014, https://www.thehenryford.org/explore/blog/clara-fords-roadside-market-a-small-building-with-big-aspirations/. For footage of Clara Ford describing her roadside market stand, see "Mrs. Ford's roadside market—outtakes," November 6, 1929, Pennsylvania Hotel, New York, NY, Fox Movietone News Story 4-80, 4 min. 9 sec., Digital Collections, University Library, University of South Carolina, Columbia, http://digital.tcl.sc.edu/cdm/ref/collection/MVTN/id/4890; "Mrs. Ford's Roadside Market—outtakes. 1929-11-20," November 4, 1929, Pennsylvania Hotel, New York, NY, Fox Movietone News Story 4-98, 1 min. 20 sec., Digital Collections, University Library, University of South Carolina, Columbia, http://digital.tcl.sc.edu/cdm/ref/collection/MVTN/id/3791; Letter to Edison Institute Schools Students about the Sale of Their School Garden Produce, 1936, available at https://www.thehenryford.org/collections-and-research/digital-collections/artifact/351141.

16. Alice Waters, *Edible Schoolyard: A Universal Idea* (San Francisco: Chronicle Books, 2008); and Edible Schoolyard Project, https://edibleschoolyard.org/ (accessed September 15, 2022).

17. Christopher Riggs and Danielle N. Lee, "Assessing Educator Perceptions of Garden-Based Learning in K–12 Science Education," *The American Biology Teacher* 84, no. 4 (2022): 213–18; and Abby M. Lohr et al., "The Impact of School Gardens on Youth Social and Emotional Learning: A Scoping Review," *Journal of Adventure Education and Outdoor Learning* 21, no. 4 (2021): 371–84. "Culturally Inclusive Teaching in the Garden," Webinar Series, KidsGardening, https://kidsgardening.org/webinar-series-culturally-inclusive-teaching/.

18. Center for Ecoliteracy in partnership with National Geographic, *Big Ideas: Linking Food, Culture, Health, and the Environment. A New Alignment with Academic Standards* (Berkeley, CA: Center for Ecoliteracy, 2014).

CHAPTER 20

A Yearbook of Science for the Public Good

David D. Vail

THE YEARBOOK of Agriculture, published annually by the US Department of Agriculture (USDA), can offer valuable contextual information to museums and historic sites as they contemplate programming options. The *Yearbook* resulted from the US government's intention to engage the public with one of the most important sectors of the US economy, agriculture. But the yearbooks meant more than that. Every volume offered a kind of science writing that related to Americans' daily lives. Topics on food, fiber, and the environment certainly had an agricultural context, but any general reader could glean interesting tips and tidbits on science and, by association, the economic, societal, and political links engendered through land practices. In this way, the yearbooks often reported on a variety of topics related to plants, animals, and soil through scientists and extension specialists. Distribution to public libraries and schools ensured that the contents reached an audience far beyond farm families. The yearbooks are a readily accessible tool to inform research and programming on a virtually unlimited number of topics.

The opening pages of the *1940 Yearbook: Farmers in a Changing World* is a good example of this versatility. Published a year before the official US entry into World War II, the *1940 Yearbook* highlighted work of agricultural scientists and argued for the heightened relevance of agriculture in a world that needed answers for a precarious political, social, and food future. USDA Secretary Henry Wallace (who was elected as vice president with President Franklin Roosevelt in 1940) expanded on the likely wartime role of agricultural science and how the yearbooks could promote scientific discoveries to expand the public knowledge for a democratic good: "One of the great solvents of passion and prejudice, which between them have pushed civilization dangerously to the brink of disaster, is the scientific spirit. I believe that on the whole this book has been written in that spirit . . . I should like to think it is a step, even if a halting one, toward that marriage of the social and the natural sciences which I believe can be one of the great contributions of democracy."[1]

Science for Field, Town, and City

These yearbooks offer data, approaches, themes, and perspectives that traverse numerous scientific disciplines. Contents give readers a sense of the blurred lines between laboratory and field work and between practical knowledge and professional expertise. Finally, the yearbooks are essentially government publications that reflect national influence but also convey era policies and politics.

Early volumes (mid- to late-nineteenth century and early twentieth century) combine chemistry, biology, plant pathology, microbiology, and conservation with the social sciences (i.e., economics, political science, and sociology). Topics range from technical experts describing concerns about farm diseases and the purity of food to food-security concerns about grain and wheat yields during wartime. Later volumes are a fascinating account of scientific debates and commentary on theme-based topics such as the uses of pesticides and agriculture's effects on the environment and climate change on farming. The *Yearbook* blends these specialties, approaches, and topics in ways that can support nearly any effort to interpret science at museums and historic sites.

A key value of the yearbooks is their ability to document the changing roles of expertise, science at land grant colleges, and the expanding networks of the schools' extension science and experiment station work year by year.[2] The annual volumes also connect the sociopolitical with the scientific and technological. The *1977 Yearbook: Gardening for Food and Fun* is a good example of a yearbook that expands on the environmentalist turn in the country and the pursuit of organic farms. Many of this volume's chapters expand on "do-it-yourself" themes for gardens in rural towns and urban communities. And the yearbooks of the 1960s and 1970s frequently address the country's counterculture influences in their own way while also tracing the agricultural science embrace of feeding the world through an agri-industrial "Green Revolution."[3]

Another way the *Yearbook of Agriculture* is a significant resource for interpreting science at museums and historic sites is how it can serve as an artifact of material culture in and of itself. Nearly every annual cover conveys an agricultural science aesthetic that can engage visitors as much as the chapter reports inside. Visitors can experience an earlier world of agricultural science through technical sketches, chapter drawings, and cover art that became defining features, especially after World War II as science became key to US dominance during the Cold War. Contributors emphasized the benefits of agricultural science discoveries, including genetic modification as conveyed in the *1986 Yearbook: Research for Tomorrow.*[4]

Just as revealing is the yearbooks' ability to engage with the USDA's problematic legacies of race, gender, class, expertise hierarchies, and environmental exploitation. Some chapters in certain years depict farmers of color in offensive language when discussing segregated extension efforts (even supporting segregation in the scientific process) or reinforce the gendered roles of women in agricultural science with surveys of home demonstration extension efforts. The *Yearbook of Agriculture* can help visitors clarify these historic complexities, as

Figure 20.1. Cover, *Yearbook of Agriculture 1986: Research for Tomorrow* (1986). The caption explains that the cover image "visualizes the hand of research working with the double helix of a molecule of DNA (deoxyribonucleic acid)—the basis of heredity in organisms—to shape tomorrow's agriculture and forestry" (n.p.). Available at https://archive.org/details/yoa1986/page/n1/mode/2up.

historian Anne Effland suggests, to better understand how "both supporters and critics of the USDA have portrayed the department as an engine of the vast changes in agriculture and food system that have taken place since the mid-nineteenth century."[5]

Records and Data

Most chapters in the early yearbooks focus on data summaries about various agricultural efforts, crop statistics, and disease outbreaks. In the *1902 Yearbook*, for example, scientists such as A. D. Hopkins (who oversaw Forest Insect Investigations with the USDA Bureau of Entomology) provided a preliminary study on "Some of the Principal Insect Enemies of Coniferous Forests in the United States." Hopkins's chapter reflects the data-driven, record-focused goals of the early yearbooks but also alludes to the fears of pathogens and pests by many experts in the same years. Also, W. H. Beal's chapter in the same year reports on the various efforts of experiment stations across the country, especially regarding agricultural science, to help local farmers.[6] In the *1903 Yearbook*, H. W. Wiley (chief of the Bureau of Chemistry and lead scientist of the Poison Squad, who would go on to also serve as first commissioner of the US Food and Drug Administration) summarized his efforts to protect consumers from adulterated foods.[7] In these ways, the *Yearbook*'s early volumes can teach visitors much about the practices of scientific knowledge creation and the trust and suspicion of expertise, both of which seemed to reinforce agency authority.

Themes and Outcomes

The USDA launched a new approach in 1936, evident in fewer statistical compilations and more theme-based analysis.[8] Most chapters follow a specific threat, agricultural issue, technology, or curiosity that relates to some of the challenges facing American agriculture of that particular year, with drought, soil erosion, and climatic changes being most urgent. The *1937 Yearbook* addressed both the Great Plains drought and how farmers could grow "better plants and animals." The *1938 Yearbook*: *Soil and Men*, however, focused on the hazards of drought and wind erosion and soil health. This shift in analytical framework came to define the scope of future volumes. Indeed, yearbooks published after World War II emphasize agricultural science for protection of resources and prevention of risks from insects and diseases and the importance of healthy fields for production goals. Throughout the 1940s and 1950s, many yearbooks stressed the key roles of knowledge and expertise in farming's future in the United States. Threats such as wind erosion, droughts, floods, and climate change and the science needed to study them are covered in detail year by year, i.e., the *1941 Yearbook: Climate and Man*.[9] Volumes published during the 1940s and 1950s also highlight agricultural science's role in wartime and in the postwar era. The yearbooks of the 1960s, 1970s, 1980s, and 1990s focused on health, wellness, and economic profit in a variety of ways that recognized local, regional, and global relationships as well as the agricultural and ecological implications of the commercial food age.[10]

Conclusion

In 1896, Assistant Secretary Charles Dabney wanted the *Yearbook of Agriculture* to be an accessible annual volume for readers to get practical advice and connect with the latest scientific advancements.[11] These volumes certainly served these goals throughout the twentieth century, and they can offer visitors an important sense of this agricultural scientific past today. The yearbooks help students, scholars, and curious visitors grasp the theories, tools, and practices that influenced our twenty-first-century foodscapes. They also offer a year-by-year material culture view of science through photographs, technical sketches, and cover art. Altogether, the yearbooks are superb examples of just the kind of artifacts that make interpreting science accessible, interesting, and useful for visitors from every walk of life.

Notes

1. US Department of Agriculture, *Yearbook of Agriculture 1940* (Washington, DC: Government Printing Office, 1940), v; and David D. Vail, "Exploring Environmental History," in *Interpreting the Environment at Museums and Historic Sites* (Lanham, MD: Rowman and Littlefield, 2019), 14–15. See also Jess Gilbert, *Agrarian Intellectuals and the Intended New Deal* (New Haven, CT: Yale University Press, 2015).
2. Vail, "Exploring Environmental History," 15. See also Debra A. Reid, *Interpreting Agriculture at Museums and Historic Sites* (Lanham, MD: Rowman & Littlefield, 2017).
3. For community farming and Green Revolution themes, see *Yearbook of Agriculture 1966: Protecting Our Food*; *Yearbook of Agriculture 1968: Science for Better Living*; *Yearbook of Agriculture 1969: Food for Us All*; *Yearbook of Agriculture 1971: A Good Life for More People*; and *Yearbook of Agriculture 1977: Gardening for Food and Fun.* For more on the history of organic farming, see Robin O'Sullivan, *American Organic: A Cultural History of Farming, Gardening, Shopping, and Eating* (Lawrence: University Press of Kansas, 2015); Andrew N. Case, *The Organic Profit: Rodale and the Making of Marketplace Environmentalism* (Seattle: University of Washington Press, 2018); and Randal S. Beeman and James A. Pritchard, *A Green and Permanent Land: Ecology and Agriculture in the Twentieth Century* (Lawrence: University Press of Kansas, 2001). See also, David D. Vail, "A Counterculture Agriculture: Organic Farming in a Commercial Food Age," in *A Companion to American Agricultural History*, R. Douglas Hurt ed. (Hoboken, NJ: Wiley Blackwell, 2022), 187–99.
4. For works on the roles of material culture and aesthetic in historical analysis, see William Thomas Okie, "Beauty and Habitation: Fredrika Bremer and the Aesthetic Imperative of Environmental History," *Environmental History* 24, no. 2 (2019): 258–81; Tim Ingold, *Being Alive: Essays on Movement, Knowledge, and Description* (London: Routledge, 2011); and W. Patrick McCray, *Making Art Work: How Cold War Engineers and Artists Forged a New Creative Culture* (Cambridge, MA: MIT Press, 2020).
5. Anne Effland, "Evolving Boundaries: 'The People's Department' Across Three Centuries," in *A Companion to American Agricultural History*, R. Douglas Hurt ed. (Hoboken, NJ: Wiley Blackwell, 2022), 258. See also Debra A. Reid, *Reaping a Greater Harvest: African Americans, the Extension Service, and Rural Reform in Jim Crow Texas* (College Station: Texas A&M University Press, 2007); Sarah Phillips, *This Land, This Nation: Conservation, Rural America, and the New Deal* (New York: Cambridge University Press, 2007); David Danbom, *The Resisted Revolution: Urban America and the Industrialization of Agriculture, 1900–1930* (Ames: Iowa State University Press, 1979); David Hamilton, *From New Day to New Deal: American Farm Policy from Hoover to Roosevelt, 1928–1933* (Chapel Hill: University of North Carolina Press, 1991); Jess Gilbert, *Planning Democracy: Agrarian Intellectuals and the Intended New Deal* (New Haven, CT: Yale University Press, 2016); Pete Daniel, *Breaking the Land: The Transformation of Cotton, Tobacco, and Rice Cultures since 1880* (Urbana: University of Illinois Press, 1986); and Pete Daniel, *Dispossession: Discrimination against African American Farmers in the Age of Civil Rights* (Chapel Hill: University of North Carolina Press, 2013).
6. See A. D. Hopkins, "Some of the Principal Insect Enemies of Coniferous Forests in the United States," in the US Department of Agriculture, *Yearbook of Agriculture 1902* (Washington, DC: Government Printing Office, 1902), 265–82. See also W. H. Beal, "Some Practical Rs of Experiment Station Work," in *Yearbook of Agriculture 1902*, 589–606.

7. See H. W. Wiley, "Determination of Effect of Preservatives in Foods on Health and Diges- tion," in the US Department of Agriculture, *Yearbook of Agriculture 1903* (Washington, DC: Government Printing Office, 1903), 289–302. See also Benjamin R. Cohen, *Pure Adulteration: Cheating on Nature in the Age of Manufactured Food* (Chicago: University of Chicago Press, 2019).

8. Henry A. Wallace, "Foreword," *Yearbook of Agriculture 1936* (Washington, DC: Government Printing Office, 1936), np.

9. See especially *Yearbook of Agriculture 1943–1947: Science in Farming*; *Yearbook of Agriculture 1953: Plant Diseases*; *Yearbook of Agriculture 1956: Animal Diseases*; *Yearbook of Agriculture 1957: Soil*; and *Yearbook of Agriculture 1958: Land*.

10. See especially *Yearbook of Agriculture 1981: Will There Be Enough Food?*; *Yearbook of Agriculture 1982: Food—From Farm to Table*; *Yearbook of Agriculture 1983: Using Our Natural Resources*; *Yearbook of Agriculture 1985: U.S. Agriculture in a Global Economy*; *Yearbook of Agriculture 1991: Agriculture and the Environment*; *Yearbook of Agriculture 1992: New Crops, New Users, New Markets*.

11. Charles W. Dabney Jr., "Preface," *Yearbook of Agriculture 1896* (Washington, DC: Government Printing Office, 1896), 4.

CHAPTER 21

Interpreting Scientists

An Interview with Storyteller Brian "Fox" Ellis

How did you get started on the path to interpreting scientists? Do you have formal training in science?

I have been a professional storyteller and science educator for more than forty years. While still in college I worked at summer camps as the resident naturalist. I have a degree in science education and have done fieldwork in ecology.[1] As a camp counselor and teacher, I found storytelling to be one of the most important tools to teach anything. At first, I told folktales that had an ecological theme. In the 1990s I was invited to portray John James Audubon for the opening of a museum exhibit. This led to more commissions to portray historical scientists. The Field Museum in Chicago invited me to portray Charles Darwin for his bicentennial, and a string of opportunities led to my becoming Meriwether Lewis, Gregor Mendel, Prince Maximillian, and most recently Robert Ridgway. I am also the communication director for Illinois Audubon.

Please describe your reasons for engaging audiences with scientists and their histories.

Through storytelling, with a few simple props, I strive to show the process of scientific discovery, for example using Audubon's paintings and ideas, his science, to introduce the audience to the integrated natural history—and ecology—of the yellow-billed cuckoo and swallowtail butterfly on a pawpaw tree. For more than thirty years, I have worked to build a longer story of scientific exploration and discovery through re-enacting the lives of several well-known scientists of the past. My goal is to engage audiences by expanding their science vocabulary and giving examples and letting them participate in the scientific process. My aim is to connect our own everyday experiences to the discipline of science. I try to create empathy with the audience, so they can see the actions of scientists as something we all do on a regular basis.

What do you mean by "science vocabulary"?

Science vocabulary emphasizes the steps of the scientific process. The first step is asking a good question that comes from observation. To test the idea, there would be an investigation that may include observing, measuring, classifying. Then, inferring based on analysis, revising or adapting the question, we would be able to communicate findings. This everyday activity—one that brings my audiences to during the field trips, is a way to understand how science develops. And, in nature study and science, we often learn more from our mistakes than we do our successes.

Your discussion of scientific methods sounds somewhat prescriptive, but science is fluid and creative. What do you read to remain current on both the history of the scientists you interpret and the methods of science (historic and contemporary) as practiced in the fields you interpret so you make the most of your discussions of the philosophy of science?

Research has always been my favorite part of presenting scientific characters. First, I read everything I can find written by the character I portray, published books and private letters or journals if I can find them. I will often read several biographies of the character as well. I then immerse myself in the time period, reading contextual history. Because I write for several science magazines, I subscribe to several others and regularly scour the Web for pertinent articles.

For example, in addition to reading Darwin's opus on *The Origin of Species* (1859), I read his more user-friendly travelogue, *The Voyage of the Beagle* (1839). I also enjoyed Irving Stone's *The Origin: A Biographical Novel of Charles Darwin* (1980). And did you know that more than ten thousand letters Darwin wrote or received are all available online thanks to the Darwin Correspondence Project? I also keep an eye out for any current research on evolution and the Galapagos Islands. Plus, I have had informal conversations with friends who are evolutionary biologists.[2]

The research can be consuming, is ongoing, but like storytelling and science it is always an adventure.

How do you take the audience along on the journey?

While I am in character, I first ask the audience to make a brief mental list of how they already use science skills in their everyday lives. When was the last time you observed something, measured something, made a prediction, or tried a new strategy to see if it would work? Sometimes I will select audience members to help me classify and organize the material relative to the work of the scientist I am interpreting. Or I may ask them to make their own predictions about outcomes I observed. This helps them realize how they think like a scientist. Incorporating opportunities for audience members to participate in the discovery process that a given scientist used makes for a longer-lasting lesson.

What are the connections between storytelling and science process skills?

I believe that storytelling models the scientific process. An audience collects information, which should prompt them to ask questions, classify details, recognize context, and predict outcomes. On the flip side, scientists are good storytellers. They communicate their findings by telling the story of their experiment. This is one of the most basic science skills. They use metaphors to convey complex ideas. They use precise language to document results.[3]

When I map out a scientist's life story that I re-create, I hit the pause button for the audience to discuss what they know so far. I create sidebars to use with selected audiences that support instructive tangents where they share their insights or ask their own questions. I pace the stories, knowing when to skim and when to dive a little deeper. This engages the audience in making predictions about outcomes. The goal is to lead audiences into the process of discovery.

Can you provide an example of storytelling techniques that embody good science process skills?

I'll share two examples!

A Storyteller's Tour—One of my favorite programs is a guided hike, usually in a wild place, a prairie, or a forest, but sometimes in a museum exhibit or urban environment. The first iteration, decades ago, was "Bird Watching with John James Audubon."[4] I then started leading hikes as other characters, like "Botanizing with Meriwether Lewis."[5] During these nature walks I quote what the scientists wrote about birds and plants, sharing their discoveries as we wander. Then I generalized the hike experience, adapting it to different settings, i.e., "Seeing the Prairie through Pioneer Eyes."

Every hike is different because it is entirely improvisational and conversational storytelling. Whatever plants or birds or soils or rocks we encounter, I guide the group in asking questions and telling stories. I always explain that we are having a conversation. I—standing in as Audubon or Lewis or a prairie resident—am not the only set of eyes and ears. I ask the audience to share what they see, hear, smell, touch, and think. For this, we usually begin with basic identification skills. "What are we seeing?" is a first step to categorize what we have identified. Is it poisonous or edible? Is it used for food or fiber? Moving toward storytelling and problem-solving, I ask: How would those who lived in the place have utilized these resources? Toward the end of the hike, I ask the audience to name the things I have identified and encourage them to tell each other stories in small groups so they are interpreting the environment for themselves based on the scientific vocabulary they have learned.

Go for a walk! First by yourself, or with a friend, and look for the science stories in your environment. As you walk through your environment, think about the stories that each object, plant, or animal could tell. How can you highlight the organic science inherent in these places? For example, for several years I have been part of the Fall Harvest Fest at The Lincoln Log Cabin State Historic Site in Lerna, Illinois, so I love to tell the story about Thomas Lincoln—Abe's father—and the fall nut harvests. Depending on what is falling from the trees, we have a conversation about his favorite wild nuts and then, with the audience, we discuss what we know about edible nuts and how to learn more about parts of the nuts used for dye.

Portraying the Historical Scientist—Because I have a full hour in a performance setting, I take my time to warm up the audience toward a more participatory engagement in science. Early in the program I ask a few rhetorical science questions to spark their inquisitive minds. When I portray Charles Darwin, I start with the big questions: "Did creation happen just once or is the story of creation ongoing? Consider all of the creatures on the earth today. Have they been here since the dawn of time? Or do new creatures arise from the old?"[6]

Later I ask a few simple fact questions everyone can answer. I collect a few responses to build confidence in participating. As John James Audubon I ask them to turn to a partner and make a list of birds that they have seen. (No wrong answers here.) I then tell the story of a few of their favorite birds. Next, I ask them to go back to their list and classify which birds are here year-round and which migrate? Again, they turn to a partner and discuss. By pacing the level of engagement and building their confidence, I can involve them in more complex science conversations by the end of the program. With Darwin we are eventually engaged in a complex conversation on the evidence for plate tectonics. With Gregor Mendel, I start with a conversation on "What traits did you inherit from your parents?" Then we build toward a complex debate on genetics and ethics.[7]

Because I portray Audubon, Darwin, and Mendel, I jumped at the chance to become ornithologist Robert Ridgway, born in Mount Carmel, Illinois, in 1850. Ridgway studied with Audubon's apprentice. Then, Ridgway was the first to apply Darwin's theory of evolution and the newly rediscovered genetics of Mendel to the field of ornithology. When I portray Ridgway, I begin with familiar backyard birds and then lead the audience in a discussion of citizen science, including backyard bird counts, and how bird feeding and the ecological restoration of their backyard can lead to long-term ecological health for birds, butterflies, and our direct descendants.

Do you incorporate modern citizen-science tools—eBird or iNaturalist or others—into your presentations? What advantages derive from doing this, in your opinion?

A resounding *Yes!* is the short answer. I have long been an advocate for citizen-science projects. And sharing eBird and iNaturalist is the one exception I regularly make to breaking character and stepping out of time. I often ask folks to open their phone and download the app while we are waiting for everyone to gather before I step into character. Because I present at bird-watching festivals, I have attended workshops with some of the folks who developed eBird. I am a huge fan. This also fulfills one of the primary goals of every program

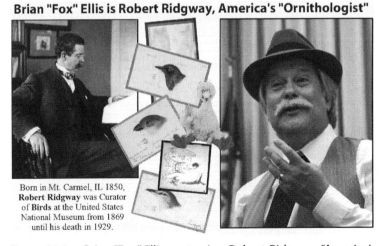

Figure 21.1.   Brian "Fox" Ellis portraying Robert Ridgway, "America's ornithologist." Promotional poster prepared by Al Harkrader. Used with permission.

I present, namely, to engage the audience in the process of science, so they are thinking like scientists, hoping they will then contribute their findings to actual scientific studies. From 100 years of Christmas Bird Counts to online eBird counts, this kind of citizen science gives us concrete data to help shape both government policy and direct actions for ecological restoration.

Do you have recommendations for historic sites or museums wanting to develop their own historic scientist programming?

We do our visitors a greater service when we engage them in reflecting on natural history, science, and its implications for our lives today. For staff at historic sites or museums wanting to do more, I suggest investigating the following questions:

What scientists lived or worked in the time periods that your historic site or museum documents?

Does your museum include any collections that document scientists?

Is your area documented in a natural history survey or a state geological survey?

Do you already interpret histories that have a science component?

Butchers, bakers, and candlestick makers all had to know animal anatomy, chemistry, and physics, respectively. Dressmaking or tailoring can lead to discussions of organic sources of cloth, with segues to geometry, math, botany, the tensile strength of two-ply threads, the engineering of water mills, and the ingenuity of the cotton gin . . . not to mention social science topics relative to politics, printing technology, international trade, industrial espionage, child-labor laws, and more. The tales of the tinker, tailor, and dressmaker all contain pathways for your own journey of discovery!

Notes

1. Brian "Fox" Ellis, *Learning from the Land: Teaching Ecology through Stories and Activities* (Englewood, CO: Teacher Ideas Press, 1997; second ed., illustrations by Vin Luong (Santa Barbara, CA: Libraries Unlimited, 2011). For more of Ellis's books, see his Amazon author page: https://www.amazon.com/stores/author/B001JP24GS/allbooks (accessed December 13, 2022).
2. The Darwin Correspondence Project, University of Cambridge, https://www.darwinproject .ac.uk/ (accessed December 13, 2022).
3. Brian "Fox" Ellis, "The Cottonwood," *Science and Children* 38, no. 4 (2001): 42–46; and Brian "Fox" Ellis, "Letters: Storytelling in Science," *Science and Children* 28, no. 5 (2001): 9.
4. Brian "Fox" Ellis, "Adventures with Audubon," Fox Tales International 2020, http://www .foxtalesint.com/index.php/20-programs/history-in-person/118-adventures-with-audubon (accessed December 10, 2022). The video can be viewed at https://youtu.be/74rH1h5849M.
5. Brian "Fox" Ellis, "Meriwether Lewis' Scientific Discoveries," one of four programs relative to "Lewis & Clark and the Corps of Scientific Discovery," Fox Tales International 2020, http://

www.foxtalesint.com/index.php/20-programs/history-in-person/121-lewis-clark-the-corps -of-discovery (accessed December 10, 2022).

6. Brian "Fox" Ellis, "Charles Darwin and his rEvolutionary Idea," Fox Tales International 2020, http://www.foxtalesint.com/index.php/20-programs/history-in-person/119-charles -darwin-and-his-revolutionary-idea (accessed December 10, 2022).

7. Brian "Fox" Ellis, "A Conversation with Gregor Mendel," Fox Tales International 2023, http://www.foxtalesint.com/index.php/20-programs/history-in-person/120-a-conversa tion-with-gregor-mendel (accessed December 10, 2022).

CHAPTER 22

Design, Science, and *Driven to Win*

A History Museum Addresses the Innovation Opportunity Gap

Rob Oleary and Matt Anderson

*RIVEN TO **Win: Racing in America**,* presented by General Motors, opened at The Henry Ford Museum of American Innovation during June 2021. This permanent exhibit explores the high-speed and flashy nature of racing by focusing on the people who designed, tested, raced, and returned to the drawing board in constant pursuit of a winner. It explores the science and technology of racing and includes race cars as material evidence of how changes in racing extended beyond the sport.

Driven to Win takes its place among a growing body of educational experiences at The Henry Ford designed to reduce the innovation opportunity gap by encouraging thought and creativity. Reducing the gap starts by engaging students in activities that require fact-finding, problem-solving, rethinking and responding, and generating a product. Reducing the gap continues by having students transfer this learning to address problems that they face. The Henry Ford's Invention Convention affords that opportunity. These involve students in group work, introduce them to innovators with proven track records, and help them engage with others who can become role models and mentors.[1]

Driven to Win reduces the innovation opportunity gap in keeping with The Henry Ford mission "to provide unique educational experiences based on authentic objects, stories, and lives from America's traditions of ingenuity, resourcefulness, and innovation. Our purpose is to inspire people to learn from these traditions to help shape a better future."[2]

The *Driven to Win Educator Resource Guide* created to pair with the *Driven to Win* exhibit includes lesson plans, an exhibit walk-through with discussion guide, and extra information on select cars called "Race Car Spotlights." Lesson plans for *Driven to Win* reiterate opportunities for creativity and for evidence-based critical thinking by stressing design while supplemental materials support educators in creating their own lessons. Lessons are differentiated for grades 3–5, 6–8, and 9–12, but they share a common theme of examining how engineers design race cars. Staying competitive in automotive racing requires near-constant optimization and design innovation. By studying the designers, the cars, and the racers who push the cars to their limits, students can learn a lot about the tension between creativity and physics, predictability and product testing, machine maintenance and human endurance. Design innovation takes these factors and more into account.

Each interdisciplinary lesson features artifacts from the collections of The Henry Ford and ends with subject-area-specific project suggestions. Students in grades 3–5 start their creative engines by pondering the question: What makes a race car go fast? As students list their ideas, they can start to categorize them under the following themes: power, weight, tire grip, heat management (and for older students, aerodynamics). Lesson prompts engage them in discovery:

Power—A car's engine supplies its power (measured in horsepower). Race cars often have large, powerful engines to get drivers around a racetrack ahead of the competition rather than the smaller engines in passenger cars designed to carry people at slower, safer speeds on a highway.

Weight—Managing the weight of the car and ensuring balance is critical in automotive design. Race cars lack things like passenger seats, which makes them lighter. Less weight translates into increased acceleration. Materials science plays a significant role here as car parts must be durable but lightweight.

Tire Grip—The friction between the tires and the road surface determines the amount of grip or traction a tire provides. This helps a car move from zero to racing speed more quickly, reduces sliding in corners, and ensures a shorter distance to stop after the driver applies the brakes.

Heat Management—Making sure components of the car are at optimal temperatures. The engine, brakes, tires, even the driver, can't be too hot or too cold to work best.

Aerodynamics—Higher grades explore racing aerodynamic goals to reduce drag as well as increase aerodynamic downforce.

Students start their analysis of cars relative to these themes by studying photographs of cars and details about components included in the curriculum materials and available via digital collections at The Henry Ford. Cars in the *Driven to Win* lesson include the 1902 Ford "999," 1935 Miller-Ford, 1956 Chrysler 300-B, 1958 Moore/Unser, 1965 Lotus-Ford, and 1984 March 84C-Cosworth. Each has photographs and specifications accessible via The Henry Ford's "Digital Collections."[3]

Students, after gathering evidence and assessing the race cars' appearances and specifications, then answer a series of questions: What determines which cars are faster than others? Does engine size mean a car with a larger engine will go faster than a vehicle with a smaller engine? Semitrucks have much larger engines than race cars but do not go as fast. Why not? Perhaps size is not the most important factor in race engine performance. What fuel do race cars use? Is it the same fuel as passenger cars? Which fuel (gasoline vs. ethanol) is best for race cars? Why?

Students can then compare race cars with passenger cars. Two cars made around the same time, the 1934 DeSoto Airflow sedan (a passenger car) and the 1935 Miller-Ford (a race car) provide an opportunity for comparison.

CAR TYPE	ENGINE	WEIGHT	POWER	TIRE GRIP	HEAT MANAGEMENT
1934 DeSoto Airflow	Inline-6	3378 pounds	Horsepower: 100 Pounds per horsepower: 33.8	Tires are half covered. More intricate tread pattern.	The grille allows air to enter and cool the engine.
1935 Miller-Ford	Ford V8	1980 pounds	Horsepower: 150 Pounds per horsepower: 13.2	Tires fully exposed. Large areas of unbroken tread pattern.	The grille allows air to enter and cool the engine.

As a group, students can identify the power of each car (horsepower). Which has more horsepower (Miller-Ford)? They first compare the weights. Which is heavier (DeSoto)? After exploring the concept of pounds relative to horsepower, students then can identify which car should be faster (fewer pounds per horsepower is faster). This illustrates one way weight is managed. Students can compare the tires on each car. What can they see about the tires (tread pattern is a complex topic but it may be interesting to note a difference)? The grilles on the front of the cars mask the heat management system. Air enters through the grilles and circulates around the engine to cool it while operating.

Who Designed the Miller-Ford?

Harry Miller was the most important American racing designer before World War II. His legacy included the four-cylinder Offenhauser engine that dominated American racing between 1935 and 1975. Miller's hallmarks were innovation, superb craftsmanship, and an artistic touch that transformed his cars, engines, and individual components into pieces of sculpture.

In 1935, Miller joined with promoter Preston Tucker in persuading Edsel Ford to return Ford Motor Company to competition for the first time since before World War I by entering Ford-powered cars in the Indianapolis 500. They formed Miller-Tucker Inc., and in a period of less than six months they built ten special race cars.

The race cars were unique.

Each car had a different color scheme. The Miller-Ford car, now in The Henry Ford collections, is blue-and-white with a large red V-8 emblem, red wheels, and leather upholstery. It illustrates all of Miller's hallmark characteristics. It is lower, sleeker, and more streamlined than any other 1935 race car. The 1935 Miller-Ford race car also featured independent front and rear suspension, unheard of in American cars of the day. Miller's race cars were the first front-drive vehicles with four-wheel independent suspension on the Indianapolis Motor Speedway, though this particular car (of the ten built) never arrived at the Speedway.

The Miller-Ford race cars incorporated 80 percent Ford components, including a stock flathead Ford V-8 engine, mounted perpendicular to the front axle to accommodate the Miller-designed front-wheel drive. The engine had a compression ratio of 9.5 to 1, produced 150 horsepower at 5,000 rpm, and was modified with special pistons, aluminum cylinder heads, an enlarged (2 gallon) oil pan, and an added 2.5-gallon reserve tank with oil cooler. A Bosch magneto operated directly from the end of the crankshaft. It also had four Stromberg dual downdraft carburetors. The tire tread was 58 inches and total automobile weight was 1,980 pounds. In addition, the Miller-Ford instrument panel included a 0–160 speedometer and gauges to measure water temperature, air pressure and oil pressure, temperature, and oil level. An air pump on the right of the dashboard was for the fuel pressure system.[4]

Each of these details resulted from engineering aimed at solving a specific challenge in race car design. Physics informed most decisions.

Did It Win?

For all its artistry, this car was not as successful as Miller's earlier designs. The hurried production schedule and lack of time for proper development led to mechanical problems. Only four of the ten cars managed to qualify for the 1935 Indianapolis 500 Speedway race, and none of those four was able to last all two hundred laps.

Figure 22.1.   1935 Miller-Ford Race Car. You can explore details, including the control panel and the aerodynamic cast-aluminum suspension arms, via THF's digital collections, available at https://www.thehenryford.org/artifact/251083/. THF90846. From the Collections of The Henry Ford.

All ten cars experienced steering troubles. The steering box was too close to the hot exhaust manifold, and when the box overheated, the gears inside expanded and locked up, causing the steering to fail. This was no disgrace to Miller, since the first twelve non-Ford finishers also boasted Miller engines and this error in design would have been easily fixed if they had had more time, but the failure soured Ford Motor Company on racing for many years afterward.[5]

Documenting the history of inventors, engineers, and their race cars, and the scientific principles at the heart of that history results in a memorable lesson. Students can transfer this knowledge to gain deeper awareness of the connections between people, innovation, and transportation. Museums and historic sites may not have flashy race cars, but they can still create lessons to engage audiences of all ages based on pounds, engine (or draft animal) size, and horsepower required to move up, down, around, and through their landscapes.

Notes

1. Invention Convention is a K–12 education program that "teaches students problem-identification, problem-solving, entrepreneurship and creativity skills, and builds confidence in invention, innovation, and entrepreneurship for life," as described at https://inhub.the henryford.org/icw/home.
2. See "Mission, Vision, and Board of Trustees," of The Henry Ford at https://www.thehenry ford.org/about/mission-and-board-of-trustees/. The two Model i frameworks of Actions of Innovation and Habits of an Innovators also influence content development. See "What Is Model i," available at https://inhub.thehenryford.org/overview/what-is-model-i.
3. Approximately 130,000 of the nearly 5,000,000 items in The Henry Ford collections are available via https://www.thehenryford.org/collections-and-research/digital-collections/.
4. You can take a closer look at the 1935 Miller-Ford Race Car at https://www.thehenryford .org/collections-and-research/digital-collections/artifact/251083. A tire tread of 58 inches equals the distance between the middle of the front two tires.
5. If you cannot visit The Henry Ford, you can read more about *Driven to Win: Racing in America*, at https://www.thehenryford.org/visit/henry-ford-museum/exhibits/driven -to-win-racing-in-america.

CHAPTER 23

Exploring Science with Young Naturalists

Debra A. Reid

NATURE STUDY provides an opportunity to learn about science, social science, and humanities. The young crabbers in the fishing community of Canarsie, Brooklyn, New York, provide an example. They lived in a place named by colonizers. They survived, as had generations of Indigenous people before them, because of knowledge gained about the flora and fauna in the sheltered harbor of Jamaica Bay. This included the Atlantic blue crab (*Callinectes sapidus*). The crabbers parlayed their knowledge into trading opportunities, informed by their knowledge of human nature based on interactions with partners, competitors, and dealers.[1]

The type of knowledge that young crabbers gained through their own reconnaissance and interactions with elders and peers was place-based and culture-bound. It developed within an ecosystem and endured because the crabbers sustained collective understanding of the blue crab within their communities and across generations. This helped them retain their ecological knowledge and preserve skills. Today we recognize this as intangible cultural heritage.[2]

Respect for science knowledge gained through experience and perpetuated through cultural engagement has been slow in coming. Two factors (at least) account for the slowness. First, access to the sources of knowledge—nature itself—declined as the natural environment changed due to urbanization, industrialization, and infrastructure expansion. Second, professionalizing the sciences privileged formal education over traditional knowledge acquisition. A growing body of scholarship makes clear, however, that traditional ecological knowledge existed before natural sciences, disciplinarily defined, existed. Recognizing that Indigenous knowledge, or native science as Gregory Cajete identifies it, resulted from observation, discovery, analysis, application, and empathy, confirms the veracity of place-based and culture-bound knowledge. This does not diminish the status of knowledge gained

Figure 23.1. Young crabbers next to an oyster dealer's wagon, Canarsie, Brooklyn, New York, 1890–1915. Photograph by Jenny Young Chandler. THF38006. From the Collections of The Henry Ford. Gift of Betty R. K. Pierce.

through avenues codified by the organizations that support professional biologists, botanists, or ecologists. Instead, it affirms traditional ecological knowledge on its own merits.[3]

Neither the young crabbers nor the Indigenous people who preceded them wrote a treatise on the Atlantic blue crab, but respect for and recognition of their process of knowledge creation now exists. A memorandum issued by the White House Office of Science and Technology Policy and the White House Council on Environmental Quality in 2021 affirms "Indigenous Knowledge as one of the many important bodies of knowledge that contributes to the scientific, technical, social, and economic advancements of the United States and to our collective understanding of the natural world," and that such knowledge will inform decision-making.[4]

Between the time when Indigenous peoples studied nature, when colonization disrupted that knowledge, and when US officials revalued that knowledge, came formalized nature study. Sally Gregory Kohlstedt credits Louis Agassi with advocating for nature study through museums and classrooms. He taught teachers in the Boston area how to implement science education as early as the 1850s. Kohlstedt recognizes the role women educators and educational administrators played in furthering the cause.[5]

Advocacy for nature study as an introduction to science education increased by 1891. Teachers at normal schools, established to train teachers, furthered the cause. Wilbur S. Jackman, a natural-science teacher at the Chicago, Illinois, Cook County Normal School, explained that "natural science, concerned largely with the earth and the living things it supports, affords the earliest and the only direct means of introducing the child to his earthly habitation . . . [and] the essence of science work . . . must be direct, individual observation." The observation laid a foundation for higher levels of thinking as a fundamentally scientific act. As Jackman explained, "true science work . . . furnishes a mental picture as a basis for reasoning [and] includes an interpretation of what has been received through the senses." With the development of this reasoning power comes "self-reliance, independence of thought, and a general strength of character." Jackman contended that science "appeals to all to become life-long students."[6]

Others created materials for teachers. One of the most well-known authors was Anna Botsford Comstock, who explained that "Nature Study is for the comprehension of the Individual life of the bird, insect or plant that is nearest at hand." She justified nature study as the means for increasing awareness of the environment for students not able to explore rural and natural places as the nation transformed from rural to urban and agrarian to industrial. She outlined the course of study, justified it, established the methodology, and field-tested it through a series of technical leaflets, compiled as the *Handbook of Nature-Study for Teachers and Parents* (1911).[7]

A public commitment to applied sciences underscored the effort to expand science education. The New York State Committee for the Promotion of Agriculture appointed Anna Comstock in 1895 to prepare a course in nature study. She introduced the course at Cornell University, a land grant institution, in 1897. Societies and government agencies had long advocated for documenting natural history generally, and insects, weeds, soils, and other geological features generally as part of support for US- and state-based economic vitality. Anna Botsford Comstock's husband, John Henry Comstock, worked as the USDA's chief entomologist before joining the faculty at Cornell University as professor of entomology and invertebrate zoology in 1882.[8] The couple proved a formidable team. John wrote, Anna illustrated, and the Comstock Publishing Company often published books that had utility within and beyond the classroom. Two books, *Manual for the Study of Insects* (1895) and *Insect Life* (1897), introduced entomology into nature study. They continued to reach scientists, too, with authoritative studies such as *How to Know the Butterflies* (1904).[9]

Others wrote books that supported other niche nature study areas. Ornithologist T. Gilbert Pearson, for instance, published *Stories of Bird Life* in 1901. He cofounded the National Association of Audubon Societies and became executive director in 1910. He launched Audubon Junior Clubs that same year. Pearson instructed teachers to collect ten cents from at least ten students, remit that to the organization, and instruct students to "learn all they can about wild birds . . . to be kind to the birds and protect them." After forming a Junior Audubon Class, teachers received buttons for each student to wear, a set of ten colored pictures, several leaflets, and the magazine, *Bird-Lore*. In exchange, teachers were "to give at least one lesson a month on the subject of birds."[10]

Public engagement in nature study helped increase awareness of natural sciences. This coincided with increased public support for preservation of forests and natural areas near growing cities and for wilderness conservation undertaken by the National Park Service and by nonprofit organizations such as the Sierra Club, founded in 1892 by John Muir. Corporations also adopted symbols of nature study to capitalize on this public enthusiasm.

Figure 23.2. Child with a butterfly net on a trade card for the Perfection Lighting Company, Grand Rapids, Michigan, 1890-1900. Photograph by Betsy Urven. Used with permission.

Figure 23.3. Alice interprets a young naturalist on a collecting expedition during the mid-nineteenth century, complete with her collecting net and display box containing snails and insects. Photograph by Betsy Urven. Used with permission.

Urban parks, forest preserves, and natural history and children's museums provided additional resources for nature study enthusiasts. Field trips proved a natural extension of the curriculum designed to get students out of doors and to open their eyes to new ways of seeing the natural world. The first children's museum in the United States provides a case in point. The Brooklyn Institute of Arts and Sciences, now the Brooklyn Museum, opened a children's museum in the Adams House, in Bedford Park (now Brower Park), on December 16, 1899. The Insect Room and the Botany Room helped guests find topics of interest. Staff installed an exhibition, *Household Insects*, in 1906 that focused on twenty common household pests. Teachers turned to the museum to supplement their nature study with museum specimens. The demand exceeded resources, however, as the museum warned school administrators not to overcrowd the lecture halls. The museum announced staff-led "field parties" and collecting trips on Saturdays during May 1907. The nature-study trips focused on butterfly and insect collecting started in June, and the museum transformed the lecture room into a laboratory where boys and girls could "learn how to preserve and mount insects." Photographs taken by Jenny Young Chandler document a model of a beetle, a photomicrograph of a flea, and children looking at nature-themed exhibits, working in the lab, and participating in a Children's Museum Nature Study Club.[11]

Formalized nature study flowered at a time often described as the nadir of race relations in the United States. National support for civil rights and social justice for all citizens waned with the Plessy v Ferguson court decision of 1896. This gave tacit approval to state and municipal legislation that challenged the notion of equal opportunity and facilitated separate and unequal educational systems. Black education occurred in separate schools that received less public funding or, if private, relied on philanthropists for funding. Black educators pursued nature study in ways that addressed the needs of Black students and their families.

George Washington Carver, scientist and manager of the experimental farm at Tuskegee Normal and Industrial Institute, published pamphlets that emphasized nature study's utility to agriculture starting in 1902. He explained in *The Need for Scientific Agriculture in the South* (1902) that all members of farm families needed to look closely at the landscape they farmed, at the "vast territory of barren and furrowed hillsides and wasted valleys," as a first step to rejuvenating the landscape that they cultivated. Being more observant of nature around them would yield evidence they needed to diagnose problems, experiment with solutions, and determine next steps. Carver recommended increased knowledge of soil science and techniques based in integrated pest management, crop rotation, and organic fertilization as reasonable next steps to take, especially in the face of boll weevil infiltration into southeastern cotton fields.[12]

Carver linked nature study to gardening in *Nature Study and Children's Games* (1904). He intended it to guide teachers who wanted to involve schoolchildren in gardening at an early age and instill in them good habits that they could take into adulthood and apply for success in an agricultural environment. After distributing all 1,500 copies of the 1904 leaflet, Carver issued a second bulletin, *Nature Study and Gardening for Rural Schools*, in 1910. Other educators called for similar integration of subject matter.[13]

Nature study remained influential for children and adults. Tuskegee Institute published *Nature's Garden for Victory and Peace* in March 1942. Credited to G. W. Carver and his assistant, Austin Curtis, *Nature's Garden* started with the poem "The Weed's Philosophy" by Martha Martin (1865–1956). It shifted immediately to the utility of "wild vegetables" as an alternative food and medicinal resource. Some entries included an illustration, some signed GWC but others probably by Carver based on botanical drawings in other publications. Each started with the common name and the botanical or scientific name of the plant and included a description of where readers, specifically those in rural Alabama in the vicinity of Tuskegee Institute, could find the plant. All included directions for preparing the plant for use, and some included recipes, both descriptive and prescriptive.[14]

What would Black crabbers of Canarsie say to the visitors to the Brooklyn Children's Museum? Would they approve of labels for the crustaceans in the museum? Who would teach whom about the crabs, or oysters, or other seafood that children in the largest city in the country ate on a regular basis? Given this context, nature study provides untapped opportunities for history museums to explore different methods of knowledge creation and validation. How-to advice drawn from historic publications can help guests compare different levels of engagement with nature, different goals of students, and different applications of the knowledge gained.

Respect for Indigenous ecological knowledge increases opportunities for museums and historical societies to engage with experts. The first Lenape-curated exhibition of Lenape cultural arts in New York City, *Lenaphoking*, opened at the Brooklyn Public Library, January 20, 2022, and ran through April 30, 2022. Providing opportunities for such cultural engagement enriches nature study.[15]

How did Indigenous populations of Canarsie manage crab fishing precontact, and post-contact? How did this compare to the young crabbers at the turn of the twentieth century? How might the students enrolled in the Children's Museum nature club have interacted with Indigenous or Black crabbers? How does what they needed to know compare to what environmentalists, park district managers, and crabbers today seek to learn about maintaining the health of Jamaica Bay and the crabs?

Close study of nature, society, and human behavior alerted generations of Canarsie crabbers to the irreparable damage that humans could do to the environment. Over time, this motivated environmentalists, including maritime experts such as Rachel Carson, to call on the public to study the environment, recognize sources of environmental degradation, and call for proactive legislation to stop the destruction. Activist nature study helped launch citizen science because it recognized the personal commitment of amateurs (e.g., noncredentialed scientists) but also recognized their role in supporting professional scientific research and advocating for change. Museums and historic sites can play a vital role in launching nature study informed by the humanities and committed to bringing all voices to the table.[16]

Notes

1. The author thanks Betsy Urven, who has interpreted, with her granddaughter, mid-nineteenth-century amateur naturalists since 2017. You can never start them too young. Betsy earned a BS from Western Illinois University with an emphasis in wildlife biology. She served on the board of the Midwest Open Air Museums Coordinating Council and was the lead interpreter and programmer at Wade House Historic site for ten years. She has a passion for living history and has volunteered at sites throughout the Midwest.

   For more on the photographer, see Cynthia Read Miller and Jeanine Head Miller, "Jenny Young Chandler: Photojournalist," blog, The Henry Ford, https://www.thehenryford .org/explore/blog/jenny-young-chandler-photojournalist; Frederick R. Black, "Jamaica Bay: A History," Gateway National Recreation Area New York, New Jersey Cultural Resource Management Study No. 3 (Washington, DC: Division of Cultural Resources, North Atlantic Regional Office, National Park Service, US Department of Interior, 1981), e-document prepared by James L. Brown (2001), available at https://www.nps.gov/parkhistory/online_books /gate/jamaica_bay_hrs.pdf; Ted Steinberg, *Gotham Unbound: The Ecological History of Greater New York* (New York: Simon & Schuster, 2014); William Wallace Tooker, comp., *Indian Names of Places in the Borough of Brooklyn, with Historical and Ethnological Notes* (New York: Francis P. Harper, 1901), Canarsie, 32–35; John A. Strong, *"We Are Still Here:" The Algonquian Peoples of Long Island Today* (Interlaken, NY: Empire State Books, 1996); John A. Strong, *The Algonquian Peoples of Long Island from Earliest Times to 1700* (Interlaken, NY: Empire State Books, 1997); and "Jamaica Bay and the Rockaways," New York City Department of

Parks and Recreation, https://www.nycgovparks.org/parks/jamaica-bay-and-the-rockaways/highlights/11918.

2. Wim Van Zanten, "Constructing New Terminology for Intangible Cultural Heritage," *Museum International* 56, nos. 1–2 (May 2004): 36–44.

3. For an overview, see Thomas F. Thornton and Shonil A. Bhagwat, eds., *The Routledge Handbook of Indigenous Environmental Knowledge* (London: Routledge, 2021); and Gregory Cajete, "Native Science and Sustaining Indigenous Communities," in *Traditional Ecological Knowledge: Learning from Indigenous Practices for Environmental Sustainability*, Melissa K. Nelson and Daniel Shilling, eds. (New York: Cambridge University Press, 2018), 15–26. For a focus on ethnobotany, see Nancy J. Turner, *Ancient Pathways, Ancestral Knowledge: Ethnobotany and Ecological Wisdom of Indigenous Peoples of Northwestern North America* (Montreal: McGill-Queen's University Press, 2014).

4. Memorandum for the Heads of Departments and Agencies from Eric S. Lander, President's Science Advisor and Director, Office of Science and Technology Policy, and Brenda Mallory, Chair, Council on Environmental Quality, Subject: Indigenous Traditional Ecological Knowledge and Federal Decision Making, November 15, 2021, https://www.whitehouse.gov/wp-content/uploads/2021/11/111521-OSTP-CEQ-ITEK-Memo.pdf.

5. Sally Gregory Kohlstedt, "Nature, Not Books: Scientists and the Origins of the Nature-Study Movement in the 1890s." *Isis* 96, no. 3 (2005): 324–52; Sally Gregory Kohlstedt, *Teaching Children Science: Hands-On Nature Study in North America, 1890–1930* (Chicago: University of Chicago Press, 2010); William T. Harris, *How to Teach Natural Science in the Public Schools* (1871), second ed. (Syracuse, NY: C. W. Bardeen, 1895), available at https://archive.org/details/howtoteachnatura00harrrich/page/n5/mode/2up.

6. Kohlstedt, "Nature, Not Books;" Kohlstedt, *Teaching Children Science*; and Wilbur S. Jackman, *Nature-Study for the Common Schools* (New York: Henry Holt and Co., 1891), quote 1–2, and 6, respectively, available at https://catalog.hathitrust.org/Record/001474972.

7. *Cornell Nature-Study Leaflets Being a Selection, with Revision, from the Teachers' Leaflets, Home Nature-Study Lessons, Junior Naturalist Monthlies and other Publications from the College of Agriculture*, Cornell University, Ithaca, NY, 1896–1904 (Albany, NY: J. B. Lyon, 1904), https://catalog.hathitrust.org/Record/006256853; Anna Botsford Comstock, *Handbook of Nature-Study for Teachers and Parents. Based on Cornell Nature-Study Leaflets, with Much Additional Material and Many New Illustrations* (Ithaca, NY: Comstock Publishing Company, 1911), quote 6.

8. Anna Botsford Comstock, *The Comstocks of Cornell: John Henry and Anna Botsford Comstock: An Autobiography by Anna Botsford Comstock*, eds. Glenn W. Herrick and Ruby Green Smith (Ithaca, NY: Cornell University Press, 1953), revised by Karen Penders St. Clair (2019); "A Guide to the John Henry and Anna Botsford Comstock Papers, 1833–1955," Cornell University, available at: http://rmc.library.cornell.edu/EAD/xml/dlxs/RMA00025.xml.

9. John Henry Comstock and Anna Botsford Comstock, *A Manual for the Study of Insects* (Ithaca, NY: Comstock Publishing Company, 1895); John Henry Comstock and Anna Botsford Comstock, *Insect Life: An Introduction to Nature Study and a Guide for Teachers, Students, and Others Interested in Outdoor Life* (New York: D. Appleton and Company, 1897); John Henry Comstock and Anna Botsford Comstock, *How to Know the Butterflies: A Manual of the Butterflies of the Eastern United States. With Forty-Five Full-Page Plates from Life Reproducing the Insects in Natural Colors* (New York: D. Appleton & Co., 1904).

10. "Junior Audubon Classes," *The Mississippi Educational Advance* 3, no. 10 (April 1914): 29.

11. For *Household Insects,* see *The Museum News (Brooklyn Institute of Arts and Sciences)* 1, no. 9 (April 1906): 137. For overcrowding during an illustrated lecture on "Seeds," see *The Museum News (Brooklyn Institute of Arts and Sciences)* 1, no. 9 (April 1906): 138. The Children's Museum maintained honey-bee hives that yielded seventy-five pounds of honey between April 15 and October 1, 1906. See *The Museum News (Brooklyn Institute of Arts and Sciences)* 2, no. 2 (November 1906): 33–34, esp. 33. For nature-study clubs, see *The Museum News (Brooklyn Institute of Arts and Sciences)* 2, no. 8 (May 1907): quote 137. Jenny Young Chandler photographs (Acquisition 32.351) include: 1) Children's Museum Insect Model, THF38134 and THF38137; 2) photomicrograph of a whole flea, THF38731, and 3) children looking at exhibits in the Bird Room, THF38138, and other nature-study themed rooms, THF38038, THF38128, THF38131, THF38133; 4) participating in a nature-study lab, THF38140; and 5) participating in a Children's Museum Nature Study Club in Bedford Park, New York, THF38136. All from the Collections of The Henry Ford. Gift of Betty R. K. Pierce.

12. Geo. W. Carver, *The Need of Scientific Agriculture in the South,* Farmers' Leaflet No. 7 (Tuskegee, AL: Tuskegee Normal and Industrial Institute, 1902); and James C. Giesen, *Boll Weevil Blues: Cotton, Myth, and Power in the American South* (Chicago: University of Chicago Press, 2011).

13. Geo. W. Carver. *Nature Study and Children's Gardens,* Teacher's Leaflet No. 2 (Tuskegee, AL: Tuskegee Normal and Industrial Institute, [1904]); and Carver, *Nature Study and Gardening for Rural Schools,* Experiment Station Bulletin No. 18 (Tuskegee, AL: Tuskegee Normal and Industrial Institute, 1910). James Ralph Jewell, *Agricultural Education including Nature Study and School Gardens,* Bulletin No. 2, 1907, Whole Number 368, Bureau of Education, Department of the Interior, second ed., rev. (Washington, DC: Government Printing Office, 1908), https://files.eric.ed.gov/fulltext/ED542841.pdf.

14. George W. Carver, *Nature's Garden for Victory and Peace* (Tuskegee, AL: Tuskegee Institute, March 1942; rev. repr. October 1942); Martha Martin, *The Weed's Philosophy and Other Poems* (n.p.: Montreal, 1913); Debra A. Reid with Deborah Evans, "What If an Artist Becomes a Scientist?" (blog), The Henry Ford, 2018, https://www.thehenryford.org/explore/stories-of-innovation/what-if/george-washington-carver.

15. Leslie, "The Borough of Homes . . . and Oysters?" Brooklyn Public Library 125th (June 24, 2010), https://www.bklynlibrary.org/blog/2010/06/24/borough-homes-and-oysters; and William W. Warner, *Beautiful Swimmers: Watermen, Crabs, and the Chesapeake Bay* (Boston: Little, Brown, and Co., 1976).

16. Trisha Gura, "Citizen Science: Amateur Experts," *Nature* 496 (2013): 259–61, https://doi.org/10.1038/nj7444-259a; and Christopher Kullenberg and Dick Kasperowski, "What Is Citizen Science?—A Scientometric Meta-Analysis," *PLOS ONE* 11 (2016): e0147152, https://doi.org/10.1371/journal.pone.0147152.

Conclusion

IDEAS: Inclusion, Diversity, Equity, and Access to Science

Karen-Beth G. Scholthof

THE INTENT of *Interpreting Science* is to explore how science themes can be developed at museums and historic sites.[1] We asked scholars to show how extant collections such as correspondence, catalogs, photographs, buildings, and their environs could be reinterpreted within the context of science during a particular historical period.[2] Museum and historic site curators, interpreters, and docents guide and educate their curious visitors. Questions could be posed to engage visitors and enrich their experiences at a museum or historic site. Scientists would lend their expertise to program planning and implementation.

The goal is to engage the public in an exploration of science. Creating a participatory experience without analytic jargon used by historians and scientists will take time, and perhaps collaboration with other experts. General questions include: Why is a particular scientific or technological advancement important? How does science influence our world? How does the world influence science? Should scientists (be able to) *explain* their science and why it is important? Can a scientist be expected to explain how and why their field of science was initiated, who performed the work, who benefited, who was ignored or not even noticed when it came time to apply the science?

Addressing Inclusion, Diversity, Equity and Access to Science (IDEAS) when identifying an object for a science-based exhibit enriches a visitor's museum experience. And, interpreting science on a local scale also helps us understand and question historical and current hierarchies of gender, race, class, sexuality, and environmentalism. How can interpreting science further social justice? A useful sociological concept is WEIRD, an acronym for

Figure 24.1. Sophie Lutterlough, Department of Entomology, National Museum of Natural History, Smithsonian Institution, Washington, D.C.. Lutterlough became the first female elevator operator at the Smithsonian Institution in 1943. She held that job for fourteen years before seeking a position at the National Museum of Natural History. Entomologist Dr. J. F. Gates Clarke hired her as an insect preparator. Over the next twenty-six years, Lutterlough took college-level courses to increase her knowledge of entomology and learned German to help with her work. She retired after forty years of service in 1983. *Source*: The Smithsonian newsletter, *TORCH*. Available at https://commons .wikimedia.org/wiki/File:Sophie_Lutterlough_at_a_Microscope,_1983.jpg.

western, educated, industrialized, rich and democratic, which can be used to draw attention to issues of social justice in science.[3]

WEIRD is typically deployed to frame differences between high-income and low-income countries. Yet, most of humanity is *not* WEIRD. We know this. How, then, do we present IDEAS for disparate audiences and how do we explore histories of those whose narratives have been excluded, exploited, or forgotten in our interpretations of science? Here, we suggest that social justice concepts can be used to interrogate WEIRD science in order to enrich an exhibit beyond reinforcing the "known" narrative of the site.[4]

Science and Social Justice

Historian Julia Rose asks, "What history is not difficult history?"[5] We entertain the parallel question: What science is not difficult (uncomfortable) science? Historians and scientists necessarily grapple with the meaning and interpretation of their research. For some scientists, decision-making may include animal or human subjects; biologicals (genetic-engineering plants, microbes, and animals), chemicals (such as pesticides and medicines), radiation (from cancer treatments to nuclear reactors), which sorts of funding are acceptable (federal agencies or multinational companies), training students or only hiring professional scientists, and working in academe or industry. Within each decision there is opportunity to consider the ramifications, then and now, for science and society. For this, inspection of the history of an object or event brings meaning to the science and its practitioners. Moreover, history frames these outcomes so that we may better understand how we arrive at our current science and how decision-making in the past can inform our future. The challenge is how do we bring in the many voices who bump up against these object narratives?

One of the first steps in this challenge is reevaluation, which is part and parcel of "doing" science. Science, like history, is framed (and contingent) on particular currents within society that provoke new strategies and ideas to solve problems, large and small. Science is a process of building, tearing down, revising, and challenging concepts. Broad questions to pique visitors' interest in an exhibit may include: What is science? Who is science for? How and why are different (and difficult) discoveries communicated to an audience? How does science connect local, regional, national, and global histories? Two strategies to consider for enveloping visitors in a particular science or technology include designated areas for reflection and visitor comments. This, in turn, can be used by curators to reshape exhibits based on what visitors understood (and enjoyed).

Therefore, (re-)interpreting science provides an ideal venue to incorporate "uncomfortable histories." Visitors arrive with their own social, cultural, economic, religious, and educational experiences. It is important that an exhibit recognize, respect, and reach out to each of these many voices in our communities and consider how they might interpret the science. Our lived experiences are complex and complicated; so too is the interpretation of a particular object and its science. Let's initiate this "interpreting science" experiment with the premise that the guest is interested in science and its impact on their daily lives and the lives of those in the past. Which aspects of the science would be interesting to communicate at a museum or historic site, who did the work, who was excluded or not seen or remembered, and who used or benefited from the science? Reading one or more chapters in this volume can be the first step in identification of objects for (re-)interpretation. In any given chapter, how has science been interpreted? Which voices come to the forefront? How would you interpret this science at your site? Can you incorporate local circumstances into a bigger science narrative? Are you comfortable being uncomfortable?[6]

Addressing the Uncomfortable

Intersectionality is a concept that may help with envisioning themes for a new(er) exhibit. Intersectionality is used to provide structure for determining an individual's place or rank in society—a rubric that reflects power and privilege.[7] Breaking down these strictures may include introducing themes of race, gender, sexuality, religion, disability, language, and education and economic status. In short, IDEAS. Most of us likely self-identify as having experienced at least one instance of marginalization related to one or more of these social strictures: how would you address this to make an object and scientific topic an exhibit theme?

Within our *Interpreting Science* framework, intersectionality invites us to look more carefully at who benefited from advances in science, who was used to advance the science, and who did or did not have a voice in determining "progress" in science. In short, who had asserted status (privilege) and power. Then, how can museums tease out the complications of science—the social construction of science, that includes: Who does the work? Who is awarded patents? Who teaches the subjects? Who can afford the price of access (time, location, economic cost) to do the science? What does science do to human relations? Class conflict? And, broad important questions: Does science relate to colonialism, and then, how can interpreting science in history museums contribute to the effort to decolonize museums?

Science with context creates a compelling, meaningful exhibit. Context is the bailiwick of the curator and historian. Science textbooks and other traditional pedagogical tools are illustrative of the lack of narrative reflecting the human experience, or more broadly construed, social justice. Science journal articles, scientists' laboratory notebooks, and conference proceedings mostly are unrevealing of personal interactions, unless one looks deeper at who is presenting or creating the materials. As we have explored in *Interpreting Science*, a wealth of material is available locally and nationally that enriches the history of science. For example, the National Tuberculosis Association developed educational films about disease and treatment intended for Black, Native American, and white communities. National Negro Health Week, a federally directed program, focused on promoting healthful living for Black Americans. In Colorado we find centuries of movement of water, first by Native Americans, then Spanish Americans, then Anglos. How did scientific advances in geology, civil engineering, ecology, and land management influence who had (and has) the water? Throughout, contributors model first steps toward looking at a local museum or historic sites to more fully dissect the who, what, when, where, and why of scientific and technological advances that shape us today.

Fluid Understandings

Science is not static or fixed—it is open to change and challenge. This should be evident as science is predicated on human thought, lived experiences, economic and political forces, and cultural imperatives. Therefore, science, like history, is advanced and enriched through ongoing interpretation and reinterpretation by experts (scientists, historians, philosophers, and social scientists) and the interested public. In fact, public engagement is an overarching, necessary aspect of social justice. For this, working within the realm of "interpreting science," it is imperative that scientists have good communication skills if there is to be epistemic exchange with the public. One mechanism to facilitate this is through conversation at museums. Many strategies can be deployed to intercalate science and social justice conversations, including formal exhibits, science cafes, hands-on activities, the arts (music, poems, drawing), and spaces for reflection. Material artifacts make science approachable and provide a creative outlet for visitors and curators. For example, postage stamps, letters, diaries, calendars, and ledger books offer a window into the ambitions of citizens and their government, reflecting perceptions of society at large.[8]

IDEAS: Inclusion, Diversity, Equity, and Access to Science

Within the context of this volume, we show that museum staff, and consulting scientists and humanists, have the wherewithal to promote and participate in greater inclusion, diversity, equity, and access to science (IDEAS). Using IDEAS we can frame the parameters by which questions about the science and its interpretation can be made: Who did the science? Who benefited from this science? Whose voices were heard or excluded? IDEAS are located within the broad framework of social justice with a tilt toward the social construction of science. That is, they illuminate the confluent influences of science, technology, and society at a particular time and place. For example, IDEAS at a science-based exhibit focused on a single object at a mid-eighteenth-century property in Philadelphia, Pennsylvania, would certainly differ from a museum of the same period with the same object in Santa Fe, New Mexico. For artifacts associated with the household, medicine, gardens, education, and agriculture, a focus on IDEAS can be used to enrich or reinterpret an exhibit.

A commonplace substance or object can be the entry into an investigation of how science was used and interpreted in the past and to reflect on the same object today. For example: tap water. Water is the single most important molecule (substance) for human life. It determines if we commit to a location, develop a community, and when to move on. Each of these aspects related to water can be explored as IDEAS. In the United States, we assume that tap water is safe to drink. Community water systems must make public annual water quality reports.[9] The science here is to locate the source of the tap water (surface water, aquifers, reservoirs, wells), where the water is treated, where untreated water is released, who does not have access to tap water: all aspects of IDEAS. Now water takes on new meaning. If you lived in Flint, Michigan, in 2015 when the alarm was sounded that lead was leaching from corroding century-old pipes, your IDEAS are based on exclusion: Water is not safe to drink, public health and city officials cannot be trusted, and you have learned that any level of lead in the blood is unsafe for children. Lead exposure in children causes neurological disorders, developmental delays, and lower IQ levels. These problems date to 1897 when the best practice was to use lead pipes as connectors to a water system. How can we interpret running water in a household through the lens of best science, medicine, and technology practices across the decades? Who lived in Flint in 1900 compared to 2015?[10]

This sort of work can extend to water sources, supplies, and water rights across the centuries. In the Rocky Mountains, the winter snowpack predicates seasonal availability of water for households, industry, and agriculture. Correspondence, diaries, newspaper archives, and historic sites can be used to study water availability (drought, floods), changing landscapes (grasslands, grazing, farming, housing development, fracking), and social justice (waterways contaminated with microbes, chemicals, slag heaps, toxic red tide). In the early to mid-twentieth century, one might investigate the technology of measuring water flow in a ditch; or reach further back to five-hundred-year-old Spanish land grants and the historic acequias that were worked for centuries; or much further back to the origins of the canals first excavated by Pueblo peoples in northern New Mexico. Or, to jump again across the centuries to today and a laboratory scientist who requires extra pure, molecular biology grade water for PCR testing of SARS-CoV-2 in human sewage to determine the relative risk of COVID-19 in a community. What is the meaning of water now?[11] Can IDEAS deepen and broaden our understanding of water and its uncomfortable history?

Conclusion

The ability to connect societal hierarchies in and of science can serve as inspiration to develop new exhibits or to revise and rethink extant exhibits. This might include crafting an exhibit to show the hierarchy of science in society (social, economic, rural, health, technology). The goal is to locate voices that have not been regarded as important or visible, through the lens of intersectionality in science. These tools, with a focus on IDEAS, will enrich the meaning and use of collections and what can be learned by curators and visitors at a museum or historic site. New voices and new perspectives will bring more meaningful interpretations and questions to the historical, social, and cultural aspects of science.

Notes

1. I thank Bethann Merkle for noting the many social justice issues addressed in *Interpreting Science*, thus providing the impetus for this chapter; and Aimee Slaughter for her elegant explanation of science as an uncomfortable history. Thoughtful and informative accounts of social justice and curating difficult histories include Julia Rose's *Interpreting Difficult Histories at Museums and Historic Sites* (Lanham, MD: Rowman & Littlefield, 2016); and Ben Railton's "Considering History: Why Learning about the Past Should Be Uncomfortable," *The Saturday Evening Post*, November 1, 2021, at https://www.saturdayeveningpost.com/2021/11/considering-history-why-learning-about-the-past-should-be-uncomfortable/.
2. Scientists are often mocked for their inability to communicate their research findings without jargon. Yet, scientists are cognizant of and are addressing social justice in the classrooms and

research environs. See Joseph L. Graves Jr., Maureen Kearney, Gilda Barabino, and Shirley Malcolm, "Inequality in Science and the Case for a New Agenda," *Proceedings of the National Academy of Sciences* 119 (2022): e2117831119. In 2015–2016, the National Academies of Science, Engineering and Medicine completed a study on "The Science of Science Communication: A Research Agenda;" the book, slides, and webinar are available at https://www.nationalacademies .org/our-work/the-science-of-science-communication-a-research-agenda#sectionCommittee.

3. Joseph Henrich, Steven J. Heine, and Ara Norenzayan, "The Weirdest People in the World?" *Behavioral and Brain Sciences* 33 (2010): 61–83. Other terms for high-income countries have included Global North, developed, and first-world. Low-income countries have been referred to as the Global South, developing, and third-world. More explicitly, I propose that in the United States, WEIRD is coded as White (ethnicity-based class system), Educated (university degrees), service-Industry (who works for whom), the Rich (or poor dichotomy), and democratic (advocacy for capitalism, wealth and success, rugged individualism, and patriotism).

4. For more on how historians can contribute to developing and contextualizing WEIRD science, see Brian Connolly, Hans Hummer, and Sara McDougall, "WEIRD Science: Incest and History," *Perspectives on History* 58, no. 5 (May 6, 2020), https://www .historians.org/research-and-publications/perspectives-on-history/may-2020/weird-science -incest-and-history.

5. Rose, *Interpreting Difficult Histories*, xiv.

6. Jajuan Johnson challenges us to be disruptive in our investigation of science, scientists, and the history of science. His contribution in this volume reveals how the Black gay male community in the United States has actively addressed HIV-AIDS with science, literature (prose and poetry), and social justice.

7. Kimberlé Williams Crenshaw, a legal scholar and Black feminist, coined and developed the concept of intersectionality. For an accessible introduction to intersectionality, see Crenshaw, "Mapping the Margins: Intersectionality, Identity Politics, and Violence Against Women of Color," in Martha Albertson Fineman, Rixanne Mykitiuk, eds., *The Public Nature of Private Violence* (New York: Routledge, 1994), 93–118, available at https://drive.google.com /file/d/1ifzT7WVGj-C7k_f0qiQDSTDxqp7bssK3/view. One example is NASA administrator Dan Goldin's quip in 1996 about "'the stale, male and pale' coterie" at the agency. The intersectionality (power and privilege) Goldin referred to at NASA was (mostly) white, male, college-educated, and economically well-situated scientists, administrators, engineers, and astronauts. The first female in space was Sally Ride (1978), and the first African American in space was Guion "Guy" Bluford (1983). For the interview, see Andrew Lawler, "Goldin Puts NASA on New Trajectory," *Science* 272 (1996): 800–802.

8. Postage stamps reflect social hubris, racial and class divides, power and privilege, and national perspective (arts, science, public health, technology, engineering, agriculture). The Smithsonian National Postal Museum has teaching resources, themed online collections, and stamps at https://postalmuseum.si.edu/search-the-collection.

9. The EPA has details on the Safe Drinking Water Act, consumer rights and water company/ utility compliance requirements and links to community water reports at https://www.epa .gov/ccr. However, demographics—ZIP codes—are the overarching and key social determinants of community health, to include water and air quality, nutrition, health care, public education, and the built environment. I thank Meg Goldberg for bringing this important point to my attention.

10. The Flint water crisis has been extensively covered by print and online media. For an in-depth analysis of this crisis in the context of social justice and IDEAS, see the "Teaching Pack: Flint, Michigan: Lethal Water" materials produced by the Global Health Education and Learning Incubator at Harvard University, at https://repository.gheli.harvard.edu/repository/collection /teaching-pack-flint-michigan-and-lethal-water/. In the United States, the tolerance level for lead in drinking water is zero (0 ppb); water with more than 15 ppb (0.015 mg/L) is "actionable," per EPA regulations (https://www.epa.gov/ground-water-and-drinking-water). The EPA describes several known outcomes associated with lead exposure: "Infants and children: Delays in physical or mental development; children could show slight deficits in attention span and learning abilities. Adults: Kidney problems; high blood pressure," https://www.epa.gov /ground-water-and-drinking-water/national-primary-drinking-water-regulations#Inorganic.

11. More on acequias can be located in this volume's chapter on irrigation history by Patricia Rettig and in Stanley G. Crawford, *Mayordomo: Chronicle of an Acequia in Northern New Mexico* (Albuquerque: University of New Mexico Press, 1988). For more on crop cultivation and Pueblo lands, a good introduction is T. Douglas Price and Anne Birgitte Gebauer, eds., *Last Hunters, First* Farmers (Santa Fe, NM: School for Advanced Research Press, 1996); and the Indian Pueblo Cultural Center (https://indianpueblo.org). Molecular biology grade ultrapure water has a maximum lead level of 10 ppb (Fisher Scientific, Catalog No. AAJ71786K2; https://www.fishersci.com/). COVID-19 is the disease caused by SARS-CoV-2 infection; for more on the US Wastewater Surveillance System, see: https://www .cdc.gov/healthywater/surveillance/; weekly data can be tracked at https://covid.cdc.gov /covid-data-tracker/#wastewater-surveillance.

Selected Readings

We compiled the following to assist those interested in interpreting science. This list is not exhaustive by any means. Instead, editors and contributors to *Interpreting Science at Museums and Historic Sites* recommend these as especially useful to museum staff (paid and unpaid) wanting to expand their knowledge of the history of science and inform their analysis and interpretation of historical resources.

Science museums collect, preserve, and interpret the history of science, and many steward artifacts and archives that can support topics that history museums can uniquely pursue. Those on the "top ten" list compiled by the National Geographic Society or *USA Today* includes museums with unparalleled collections as well as relatively new science centers. Reach out to curatorial and programming staff at institutions near you to discuss the potential of collaborative work to deepen public engagement. You can start by perusing collections at science museums recognized for their historical collections. These include the Science History Institute (Philadelphia, Pennsylvania), the Franklin Institute (Philadelphia, Pennsylvania), the Smithsonian Institution's National Museum of Natural History and National Air and Space Museum (Washington, D.C.), the Museum of Science and Industry (Chicago, Illinois), and the Bishop Museum (Honolulu, Hawai'i), to name a few. Many offer provocative overviews of scientific collections relative to current topics, including social, cultural, and historical context. See, among others, the Wellcome Collection, London, "a free museum and library exploring health and human experience" (https://wellcomecollection.org/).

Science: History, Meaning, and Method

General readings in the history of science are numerous, and summations of key topics and themes abound. Some include *The Cambridge History of Science*, 8 vols. (Cambridge, UK:

Cambridge University Press, 2018–2020); Peter Bowler and Iwan Morus, *Modern Science: A Historical Survey* (Chicago: University of Chicago Press, 2005); Charles C. Gillispie, *Dictionary of Scientific Biography*, 16 vols. (New York: Scribners, 1970–1980); Steven Shapin, *The Scientific Revolution* (Chicago: University of Chicago Press, 1996); Lawrence M. Principe, *The Scientific Revolution: A Very Short Introduction* (New York: Oxford University Press, 2011); and John Waller, *Fabulous Science: Fact and Fiction in the History of Scientific Discovery* (Oxford: Oxford University Press, 2002).

Others explore doubt and the role of scientific meaning and method relative to doubt and popular acceptance or rejection of science. Two key works stand out: Erik M. Conway and Naomi Oreskes, *Merchants of Doubt: How a Handful of Scientists Obscured the Truth on Issues from Tobacco Smoke to Global Warming* (New York: Bloomsbury Press [repr. ed.], 2010); and Naomi Oreskes, *Why Trust Science? Why the Social Character of Scientific Knowledge Makes It Trustworthy* (Princeton, NJ: Princeton University Press, 2021). Both of these books will serve readers of *Interpreting Science* well as they consider the archives and material culture artifacts documenting doubt, suspicion, and cultural rejection of scientific expertise.

Other general works survey scientific meaning and methods and its role in medicine in specific centuries. On science, medicine, and practice, see W. F. Bynum, *Science and the Practice of Medicine in the Nineteenth Century* (Cambridge: Cambridge University Press, 1994); John Harley Warner, ed., *Locating Medical History: The Stories and Their Meanings* (Baltimore: Johns Hopkins University Press, 2004); John V. Pickstone, ed., *Medical Innovations in Historical Perspective* (Basingstoke, Hampshire: Macmillan, 1992); Andrew Pickering, *The Mangle of Practice: Time, Agency, and Science* (Chicago: University of Chicago Press, 1995); Karin Knorr Cetina, Theodore R. Schatzki, and Eike von Savigny, eds., *The Practice Turn in Contemporary Theory* (London: Routledge, 2000); and Andrew Cunningham and Perry Williams, *The Laboratory Revolution in Medicine* (Cambridge: Cambridge University Press, 1992). Additional works on "ways of knowing" could include Terri M. Romano, *Making Medicine Scientific: John Burdon Sanderson and the Culture of Victorian Science* (Baltimore: Johns Hopkins University Press, 2002); and John V. Pickstone, *Ways of Knowing: A New History of Science, Technology, and Medicine* (Manchester: Manchester University Press, 2000; Chicago: University of Chicago Press, 2001).

Of course, academic journals (often behind membership paywalls) have equal importance in providing general overviews on a variety of scientific topics. Some of the central publications include *Isis*, *Osiris*, and *Technology and Culture*. For work related to the history of science in agricultural and environmental context, see *Agricultural History* and *Environmental History*.

Science: Knowledge and Expertise

Another key dimension (and tension) in the history of science includes knowledge and expertise. Key works on each and the complicated dynamic between both include Tim Ingold, *Being Alive: Essays on Movement, Knowledge, and Description* (London: Routledge, 2011); Robert E. Kohler, *Landscapes and Labscapes: Exploring the Lab-Field Border in Biology* (Chicago: University of Chicago Press, 2002); Michael Canfield, eds., *Field Notes on Science* and *Nature* (Cambridge, MA: Harvard University Press, 2011); and Katherine Pandora, "Knowledge Held in Common: Tales in Luther Burbank and Science in the American Vernacular," *Isis* 92, no. 3 (2001): 484–516. The costs of limiting access to knowledge based on gender or race has received attention from Amy Slaton, *Race, Rigor and Selectivity in U.S. Engineering: The History of an Occupational Color Line* (Cambridge, MA: Harvard University Press, 2010). See also "Women in Science."

On the relationships between environmental history and the history of science, see Mark D. Hersey and Jeremy Vetter, "Shared Ground: Between Environmental History and the History of Science," *History of Science* 57, no. 4 (2019): 403–40; and Dolly Jørgensen, Finn Arne Jørgensen, and Sara B. Pritchard, *New Natures: Joining Environmental History with Science and Technology Studies* (Pittsburgh: University of Pittsburgh Press, 2013).

Antibiotics

The theme of antibiotics helps practitioners connect a variety of themes related to the history of science (human, nonhuman, environment, medicine). On agricultural science and antibiotics, see Orville Schell, *Modern Meat: Antibiotics, Hormones, and the Pharmaceutical Farm* (New York: Random House, 1984); Ted Genoways, *The Chain: Farm, Factory, and the Fate of Our Food* (New York: Harper, 2014); and Maryn McKenna, *Big Chicken: The Incredible Story of How Antibiotics Created Modern Agriculture and Changed the Way the World Eats* (Washington, DC: National Geographic, 2017). For more generalized works that are more specific to eras, see Kendra Smith-Howard, *Pure and Modern Milk: An Environmental History since 1900* (New York: Oxford University Pres, 2014); and Marc Landas, *Cold War Resistance: The International Struggle over Antibiotics* (Lincoln: University of Nebraska Press, 2020).

Astronomy/Astrophysics

A familiar topic in the long history of science but which has often carried a popular, public interest is the histories of astronomy and astrophysics. On accessible works that could serve historic sites and museums well, see David Baron's *American Eclipse: A Nation's Epic Race to Catch the Shadow of the Moon and Win the Glory of the World* (New York: W. W. Norton, 2017); Dava Sobel, *The Glass Universe: How the Ladies of the Harvard Observatory Took the Measure of the Stars* (New York: Viking Press, 2017); Neil deGrasse Tyson, *Astrophysics for People in a Hurry* (New York: W. W. Norton, 2017); Michael Hoskin, *History of Astronomy: A Very Short Introduction* (Oxford: Oxford University Press, 2003); Brian Greene, *The Fabric of the Cosmos: Space, Time, and the Texture of Reality* (New York: Random House, 2004); and W. Patrick McCray, *Giant Telescopes: Astronomical Ambition and the Promise of Technology* (Cambridge, MA: Harvard University Press, 2006).

Additional works that link histories of scientific revolution to astronomy include Dennis Richard Danielson, *The First Copernican: Georg Joachim Rheticus and the Rise of the Copernican Revolution* (New York: Walker, 2006); Maurice A. Finocchiaro, *The Galileo Affair: A*

Documentary History (Berkeley: University of California Press, 1989); Peter Machamer, ed., *The Cambridge Companion to Galileo* (Cambridge: Cambridge University Press, 1998); David H. DeVorkin, ed., *The History of Modern Astronomy and Astrophysics: A Selected, Annotated Bibliography* (New York: Garland, 1982); and Carroll W. Pursell Jr., *Astronomy in America* (Chicago: Rand McNally, 1967).

Atomic Science

There are many different historical threads to consider when it comes to the history of the atom, atomic science, and nuclear technologies. For more general works that address these themes and how they interconnect, see Angela N. H. Creager, *Life Atomic: A History of Radioisotopes in Science and Medicine* (Chicago: University of Chicago Press, 2013); Adam Higginbotham, *Midnight in Chernobyl: The Untold Story of the World's Greatest Nuclear Disaster* (New York: Simon & Schuster, 2019); and Timothy Jorgensen, *Strange Glow: The Story of Radiation* (Princeton, NJ: Princeton University Press, 2016). For the geopolitics of atomic science, World War II, and the Cold War, see Jacob Darwin Hamblin, *The Wretched Atom: America's Global Gamble with Peaceful Nuclear Technology* (New York: Oxford University Press, 2021); Alex Wellerstein, *Restricted Data: The History of Nuclear Secrecy in the United States* (Chicago: University of Chicago Press, 2021); and Kate Brown, *Plutopia: Nuclear Families, Atomic Cities, and the Great Soviet and American Plutonium Disasters* (New York: Oxford University Press, 2013). For topics related to environments, testing, and the socio-scientific consequences, see Sarah A. Fox, *Downwind: A People's History of the Nuclear West* (Lincoln, NE: Bison Books, 2014); Andy Kirk, *Doom Towns: The People and Landscapes of Atomic Testing* (New York: Oxford University Press, 2016); and Neil Oatsvall, *Atomic Environments: Nuclear Technologies, the Natural World, and Policymaking, 1945–1960* (Tuscaloosa: University of Alabama Press, 2023).

Agricultural Science

The integration of agricultural sciences through archives, artifacts, and historical works offers accessible and interactive sources for museums and historic sites to interpret science in a variety of ways for a variety of crowds. For general histories on practices, policies, and places related to agricultural science, see Anne Effland, "Evolving Boundaries: 'The People's Department' Across Three Centuries," *A Companion to American Agricultural History*, R. Douglas Hurt, ed. (Hoboken, NJ: Wiley Blackwell, 2022); Benjamin R. Cohen, *Notes from the Ground: Science, Soil, and Society in the American Countryside* (New Haven, CT: Yale University Press, 2009); and Emily Pawley, *Nature of the Future: Agriculture, Science, and Capitalism in the Antebellum North* (Chicago: University of Chicago Press, 2020). For works that explore agricultural experiment stations, extension work, and land grant universities, see Alan I. Marcus, *Science as Service: Establishing and Reformulating American Land-Grant Universities, 1865–1930* (Tuscaloosa: University of Alabama Press, 2015); and Marcus, *Service as Mandate: How American Land-Grant Universities Shaped the Modern World, 1920–2015*

(Tuscaloosa: University of Alabama Press, 2015). See also, David Danbom, *Our Purpose Is to Serve: The First Century of the North Dakota Agricultural Experiment Station* (Fargo: North Dakota Institute for Regional Studies, 1990); and David D. Vail, "Demonstration Trains and the Rise of Mobile Agricultural Science in the Great Plains," *Great Plains Quarterly* 38 (Spring 2018): 151–74.

On agricultural science, plants, and a move toward global food production, see Charles C. Mann, *The Wizard and the Prophet: Two Remarkable Scientists and Their Dueling Visions to Shape Tomorrow's World* (New York: Alfred A. Knopf, 2018); R. Douglas Hurt, *The Green Revolution in the Global South: Science, Politics, and Unintended Consequences* (Tuscaloosa: University of Alabama Press, 2018); Christine Keiner, *The Oyster Question: Scientists, Watermen, and the Maryland Chesapeake Bay since 1880* (Athens: University of Georgia Press, 2010); Noel Kingsbury, *Hybrid: The History of Science of Plant Breeding* (Chicago: University of Chicago Press, 2011); Helen Anne Curry, *Evolution Made to Order: Plant Breeding and Technological Innovation in Twentieth-Century America* (Chicago: University of Chicago Press, 2016); and Barbara Kingsolver, *Animal, Vegetable, Miracle: A Year of Food Life* (New York: HarperCollins, 2007).

Art and Science

Engaging approaches in interpreting science for visitors to museums and historic sites might include the blending of art and science to reveal the interconnections between STEM and Humanities. Key works that help with the conceptualization, framework, and documentation of sources related to art and science include W. Patrick McCray, *Making Art Work: How Cold War Engineers and Artists Forged a New Creative Culture* (Cambridge, MA: MIT Press, 2020); Bruno Lator, *Science in Action: How to Follow Scientists and Engineers Through Society* (Cambridge, MA: Harvard University Press, 1988); Travis Nygard, "George Washington Carver (ca. 1863–1943)," in *Unforgettables: An Alternate History of American Art*, Charles Eldredge, ed. (Oakland: University of California Press, 2022), 283–88; and Debra A. Reid with Deborah Evans, "What If an Artist Becomes a Scientist?" *The Henry Ford* [2018], https://www.thehenryford.org/explore/stories-of-innovation/what-if/george-washington-carver.

Chemistry

Chemistry is embedded in our lives. Elements and compounds, organic, inorganic, and synthetic, take many forms, including water, leather, alcohol, fertilizers, gasoline, jewelry, pharmaceuticals, and textiles. There are good, general works that open the door to chemistry by exploring everyday objects. Scholarly work, from chemistry journals, textbooks, and historical analyses, provide broader and deeper insights into the subject matter. The American Chemical Society (https://www.acs.org/) has interesting and accessible materials for educators and the public. Books that provide a generalist's entry point to chemistry include Mai Thi Nguyen-Kim, *Chemistry for Breakfast: The Amazing Science of Everyday Life* (Vancouver,

Canada: Greystone Books, 2021); Mark Miodownik, *Stuff Matters: Exploring the Marvelous Materials That Shape Our Man-Made World* (New York: Viking, 2014); Mark Miodownik, *Liquid: The Delightful and Dangerous Substances That Flow Through Our Lives* (New York: Viking, 2018); George Zaidan, *Ingredients: The Strange Chemistry of What We Put in Us and on Us* (New York: Dutton, 2020); Sam Kean, *The Disappearing Spoon: And Other True Tales of Madness, Love, and the History of the World from the Periodic Table of the Elements* (New York: Little Brown, 2010); Simon Garfield, *Mauve: How One Man Invented a Color That Changed the World* (New York: W. W. Norton, 2000); J. Kenji López-Alt, *The Food Lab: Better Home Cooking Through Science* (New York: W. W. Norton, 2015); Deborah Blum, *The Poisoner's Handbook: Murder and the Birth of Forensic Medicine in Jazz Age New York* (New York: Penguin, 2011); Frank A. Von Hippel, *The Chemical Age: How Chemists Fought Famine and Disease, Killed Millions, and Changed Our Relationship with the Earth* (Chicago: University of Chicago Press, 2020); Paul Strathern, *Mendeleyev's Dream: The Quest for the Elements* (New York: Pegasus, 2019); and Philip Ball, *The Elements: A Visual History of Their Discovery* (Chicago: University of Chicago Press, 2021).

Scholarly historiography of chemical sciences includes Victoria Lee, *The Arts of the Microbial World: Fermentation Science in Twentieth-Century Japan* (Chicago: University of Chicago Press, 2021); and Benjamin Johnson, *Making Ammonia: Fritz Haber, Walther Nernst, and the Nature of Scientific Discovery* (Cham, Switzerland: Springer Nature, 2022), an open-access book available at https://link.springer.com/book/10.1007/978-3-030-85532-1. See also "Pesticides, Toxicology, and Rachel Carson."

Chemists assess their craft, indicating the range of subjects that affect their worldview, i.e., *Roald Hoffman on the Philosophy, Art, and Science of Chemistry*, Jeffrey Kovac and Michael Weisberg, eds. (New York: Oxford University Press, 2012). Oliver Sacks's first autobiography, *Uncle Tungsten: Memories of a Chemical Boyhood* (New York: Vintage Books, 2001), combined inspiration from his extended family along with a history and science of chemistry.

Climate Change

For an overview of climate change related to public history and memory, see David Glassberg, "Place, Memory, and Climate Change," *The Public Historian* 36, no. 3 (2014): 17–30. For accessible, general readings, see Katrin Kleemann and Jeroen Oomen, eds., "Communicating the Climate: From Knowing Change to Changing Knowledge," *RCC Perspectives: Transformations in Environment and Society*, no. 4 (2019): 1–122. For a comprehensive overview of relevant topics and sources, see Sam White, Christian Pfister, and Franz Mauelshagen, eds., *The Palgrave Handbook of Climate History* (London: Palgrave Macmillan UK, 2018). Other general works include Bruno Latour, *Down to Earth: Politics in the New Climatic Regime* (Cambridge: Polity, 2018); and Jared Farmer, *Elderflora: A Modern History of Ancient Trees* (New York: Basic Books, 2022).

Many see climate change as a consequence of human activity so disruptive that it has arguably resulted in a new epoch in geological time, the Anthropocene, or the Age of Man. For two thought-provoking overviews, see J. R. McNeill and Peter Engelke, *The Great Acceleration: An Environmental History of the Anthropocene* (Cambridge, MA: Belknap Press of Harvard University Press, 2014) and Julia Adeney Thomas, Mark Williams, and Jan Zalasiewicz, *The Anthropocene: A Multidisciplinary Approach* (Cambridge, UK: Polity, 2020).

Diseases and Microbes

The interpretation of science related to microbes and diseases comes from a variety of scholarly efforts that blend the histories of science and disease, the history of medical science, agricultural history, environmental history, and even military history to name a few. The following works try to identify these emphases in useful ways. For medical bacteriology and germ theory, see Christoph Gradmann, *Laboratory Disease: Robert Koch's Medical Bacteriology* (Baltimore: Johns Hopkins University Press, 2009); Gerald L. Geison, *The Private Science of Louis Pasteur* (Princeton, NJ: Princeton University Press, 1995); and Michael Worboys, *Spreading Germs: Disease Theories and Medical Practice in Britain, 1865–1900* (Cambridge: Cambridge University Press, 2000).

On specific diseases, especially epidemics, and their socio-cultural-economic-political implications, see Charles E. Rosenberg, *The Cholera Years: The United States in 1832, 1849, and 1866* (Chicago: University of Chicago Press, 1987); Charles E. Rosenberg, *Explaining Epidemics and Other Studies in the History of Medicine* (Cambridge: Cambridge University Press, 2010); and Frank M. Snowden, *Epidemics and Society: From the Black Death to the Present* (New Haven: Yale University Press, 2019). On tuberculosis, see Helen Bynum, *Spitting Blood: The History of Tuberculosis* (Oxford: Oxford University Press, 2012); Thomas Dormandy, *The White Death: A History of Tuberculosis* (New York: New York University Press, 2000); and Katherine Ott, *Fevered Lives: Tuberculosis in American Culture Since 1870* (Cambridge, MA: Harvard University Press, 1996).

The home as a vector galvanized research into diseases associated with and transmitted by close human contact, often in domestic situations. See Nancy Tomes, *The Gospel of Germs: Men, Women, and the Microbe in American Life* (Cambridge, MA: Harvard University Press, 1998); and Judith Walzer Leavitt, *Typhoid Mary: Captive to the Public's Health* (Boston: Beacon Press, 1996).

For works that survey zoonotic exchanges and human-animal-plant relationships, see E. Fuller Torrey and Robert H. Yolken, *Beasts of the Earth: Animals, Humans, and Disease* (New Brunswick, NJ: Rutgers University Press, 2005); Alan L. Olmstead and Paul Webb Rhode, *Arresting Contagion: Science, Policy, and Conflicts over Animal Disease Control* (Cambridge, MA: Harvard University Press, 2015); Robert Wallace, *Big Farms Make Big Flu: Dispatches on Infectious Disease, Agribusiness, and the Nature of Science* (New York: Monthly Review Press, 2016); and Ed Yong, *I Contain Multitudes: The Microbes within Us and a Grander View of Life* (New York: HarperCollins, 2016).

DNA

DNA history offers much for museums and historic sites to consider when building exhibits and recommending collections for researchers and the general public alike. Some selected general readings on this topic include Angela N. H. Creager, "Recipes for Recombining DNA: A History of Molecular Cloning: A Laboratory Manual," *BJHS Themes* 5 (2020): 225–43; Siddhartha Mukherjee, *The Gene: An Intimate History* (New York: Scribner, 2016); and Brenda Maddox, *Rosalind Franklin: The Dark Lady of DNA* (New York: Harper Perennial, 2003). Another resource to consider is the PBS documentary *American Masters: Decoding Watson*, which focused on geneticist and biologist James Watson's key role in DNA research, including his co-discovery of DNA structure.

Domestic Science/Home Economics

Male-dominated social constructions marginalized women in many ways. Joan Jacobs Brumberg and Nancy Tomes explain how this limited female participation in science and medicine, among other disciplines, launched a feminized service profession (domestic science or home economics), in "Women in the Professions: A Research Agenda for American Historians," *Reviews in American History* 10, no. 2 (June 1982): 275–96. Sarah Stage and Virginia Bramble Vincenti compiled essays that address the subject to 1994 with the renaming of home economics as "Family and Consumer Sciences," in *Rethinking Home Economics: Women and the History of a Profession* (Ithaca, NY: Cornell University Press, 1997). Women trained in home economics, and men who facilitated their work applied their formal training in nutrition, hygiene, public health, child development, and family economics in many ways. They influenced personal and familial decision-making, informed curriculum at schools, and influenced community engagement undertaken by churches and benevolent associations, as well as local, state, and national policy. At Cornell University, students, archivists, and faculty collaborated on a virtual exhibit, *From Domesticity to Modernity: What Was Home Economics?* available at https://rmc.library.cornell.edu/homeEc /default.html, and facilitated access to hundreds of publications from the 1850s to the 1950s, consolidated in the digital database *HEARTH: Home Economics Archive: Research, Tradition, History*, available at https://digital.library.cornell.edu/collections/hearth. Published sources such as this can provide a solid foundation on which to build interpretation around science in domestic life, about middle-class female professional identity and activism, and about community-based public health initiatives.

Home economics also informs interpretation of gender and race within a marginalized science. Black and white women engaged in this work, much conducted through classrooms at segregated schools and state-based home demonstration extension programs. Across the United States, this work operated out of land grant colleges created as a result of the 1862 Morrill Land-Grant Act. In former Confederate states, two colleges, the 1862 land grant college serving white students and the 1890 Morrill land grant institution serving Black students, operated separate programs until the early 1970s, after which national legislation directed 1890 institutions to serve economically disadvantaged and ethnically diverse populations through extension programs. Studies tend to focus on either white women or black women or a state or region, as does Lu Ann Jones, *Mama Learned Us to Work: Farm Women in the New South* (Chapel Hill: University of North Carolina Press, 2002); and Debra A. Reid, *Reaping a Greater Harvest: African Americans, the Extension Service, and Rural Reform in Jim Crow Texas* (College Station: Texas A&M University Press, 2007), but the science at the heart of the training and the programming rarely receives attention. For a blog exploring the repercussions of racism, sexism, nutrition, and public health, see the blog, Debra A Reid, "Food Soldiers: Nutrition and Race Activism," The Henry Ford, February 10, 2021, available at https://www.thehenryford.org/explore/blog/food-soldiers-nutrition-and-race-activism.

Electricity

For the topic of electricity, museums and historic sites have numerous works to consult and consider. For a good general history, see James Delbourgo, *A Most Amazing Scene of Wonders: Electricity and Enlightenment in Early America* (Cambridge, MA: Harvard University Press, 2006); David Bodanis, *Electric Universe: How Electricity Switched on the Modern World* (Portland, OR: Broadway Books, 2006); Thomas P. Hugues, *Networks of Power: Electrification in Western Society, 1880–1930* (Baltimore: Johns Hopkins University Press, 1993); and Linda Simon, *Dark Light: Electricity and Anxiety from the Telegraph to the X-Ray* (New York: Houghton Mifflin Harcourt, 2004). Inventors such as Thomas Edison, Nikola Tesla, and George Westinghouse also serve as key historical actors in presenting this past. W. Bernard Carlson, *Tesla: Inventor of the Electrical Age* (Princeton, NJ: Princeton University Press, 2015); and Jill Jones, *Empire of Light: Edison, Tesla, Westinghouse, and the Race to Electrify the World* (New York: Random House, 2003) provide accessible histories for curators and docents to use in their interpretive efforts. On region and electricity, see Leah S. Glaser, *Electrifying the Rural American West: Stories of Power, People, and Place* (Lincoln: University of Nebraska Press, 2009).

Environmental Science

The history of environmental science supports museum interpretation in numerous ways. For works that address social and cultural environmentalism, see Leah Penniman, *Black Earth Wisdom: Soulful Conversations with Black Environmentalists* (New York: Amistad, 2023); and Erin Sharkey, ed., *A Darker Wilderness: Black Nature Writing from Soil to Stars* (Minneapolis: Milkweed Editions, 2023). On the role of expertise and authority relative to the environment, see Stephen Bocking, *Nature's Experts: Science, Politics, and the Environment* (New Brunswick, NJ: Rutgers University Press, 1974); Samuel P. Hays, *Beauty, Health, and Permanence: Environmental Politics in the United States, 1955–1985* (New York: Cambridge University Press, 1987); and Jeremy Vetter, *Field Life: Science in the American West during the Railroad Era* (Pittsburgh: University of Pittsburgh Press, 2016). For the need to strike a balance between agriculture and environmental stewardship, see Randal S. Beeman and James A. Pritchard, *A Green and Permanent Land: Ecology and Agriculture in the Twentieth Century* (Lawrence: University Press of Kansas, 2001); and Sheldon Krimsky and Roger P. Wrubel, *Agricultural Biotechnology and the Environment: Science, Policy, and Social Issues* (Urbana: University of Illinois Press, 1996).

Evolution

There are numerous works on the history of evolution—the specifics on theory, intellectual debates, philosophical quandaries, and real-world applications. For a good primer work on the history of Charles Darwin's scholarly work, including evolutionary theory, see Duncan M. Porter and Peter W. Graham, eds., *The Portable Darwin* (New York: Penguin Classics, 1993). For philosophical debates around evolution, see Michael Ruse, *Mysteries of Mysteries: Is Evolution a Social Construction?* (Cambridge, MA: Harvard University Press, 1999); and Michael Ruse, *Darwin and Design: Does Evolution Have a Purpose?* (Cambridge, MA: Harvard University Press, 2003). For Darwin's intellectual legacies, see Devin Griffiths and Deanna Kreisel, eds., *After Darwin: Literature, Theory, and Criticism in the Twenty-First Century* (Cambridge: Cambridge University Press, 2022). On the relationship between evolution, ecology, and evolutionary history, see Sharon E. Kingsland, *The Evolution of American Ecology, 1890–2000* (Baltimore: Johns Hopkins University Press, 2005); Susan R. Schrepfer and Philip Scranton, eds., *Industrializing Organisms: Introducing Evolutionary History* (New York: Routledge, 2004); and Edmund Russell, *Uniting History and Biology to Understand Life on Earth* (Cambridge: Cambridge University Press, 2011).

Food Safety

The topic of food in general offers museums and historic sites many opportunities to address science topics. See, for example, Marion Nestle, *Unsavory Truth: How Food Companies Skew the Science of What We Eat* (New York: Basic Books, 2018). For concepts that incorporate histories of poisons, food purity, and related policies, see Deborah Blum, *The Poison Squad: One Chemist's Single-Minded Crusade for Food Safety at the Turn of the Twentieth Century* (New York: Penguin Press, 2018); Benjamin R. Cohen, *Pure Adulteration: Cheating on Nature in the Age of Manufactured Food* (Chicago: University of Chicago Press, 2019); Clayton A. Coppin and Jack High, *The Politics of Purity: Harvey Washington Wiley and the Origins of Federal Food Policy* (Ann Arbor: University of Michigan Press, 1999); Lorine Swainston Goodwin, *The Pure Food, Drink, and Drug Crusaders, 1879–1914* (Jefferson, NC: McFarland, 1999); and James Harvey Young, *Pure Food: Securing the Federal Food and Drugs Act of 1906* (Princeton, NJ: Princeton University Press, 1989).

Garden Schools/School Gardens

Agricultural and educational reformers, intent on engaging youth, embraced school gardens as a medium to teach about natural and agricultural science and food production. A rich literature published between the 1870s and the present affirm sustained commitment to school gardens, now considered foundational to furthering the edible education initiative launched by chef and restaurateur Alice Waters.

Surveys of published resources include Emily Marsh, "The School Garden," National Agricultural Library (n.d.), https://www.nal.usda.gov/collections/stories/school

-gardens; Constance Carter, "School Gardens," Webcast (2007), https://www.loc.gov/item/2021687884/; and Constance Carter, "School Gardens," Library of Congress Websites of Interest (2015), https://www.loc.gov/rr/program/journey/schoolgardens.html.

US government technical bulletins included are three by George Washington Carver: *Progressive Nature Study* (Tuskegee, AL: Tuskegee Normal and Industrial Institute, 1897); *Nature Study and Children's Gardens*, Teacher's Leaflet No. 2 (Tuskegee, AL: Tuskegee Normal and Industrial Institute, 1904); and *Nature Study and Gardening for Rural Schools*, Experiment Station Bulletin No. 18 (Tuskegee, AL: Tuskegee Normal and Industrial Institute, 1910). The US Department of Agriculture promoted gardens in several bulletins and how this related to a normal school (or teacher college) curriculum. See, for example, Beverly Thomas Galloway, *School Gardens: A Report upon Some Cooperative Work with the Normal Schools of Washington, with Notes on School-Garden Methods Followed in Other American Cities*, US Department of Agriculture, Office of Experiment Stations, Bulletin No. 160 (Washington, DC: Government Printing Office, 1905); James Ralph Jewell, *Agricultural Education including Nature Study and School Gardens*, Bulletin No. 2, 1907, Whole Number 368, Bureau of Education, Department of the Interior, second ed., rev. (Washington, DC: Government Printing Office, 1908); F. W. Howe, *Boys' and Girls' Agricultural Clubs*, US Department of Agriculture, *Farmers' Bulletin* 385 (Washington, DC: Government Printing Office, 1910); and Susan Bender Sipe, *Some Types of Children's Garden Work*, US Department of Agriculture, Office of Experiment Stations, Bulletin Number 252 (Washington, DC: Government Printing Office, 1912).

School garden advocates also found outlets through popular presses. These include in order of publication: W. S. Jackman, *Nature-Study for the Common Schools* (New York: Henry Holt and Company, 1891); Louise Klein Miller, *Children's Gardens for School and Home: A Manual of Cooperative Gardening* (New York: D. Appleton and Company, 1904); Clarence Moores Weed and Philip Emerson, *The School Garden Book* (New York: Charles Scribner's Sons, 1909); Mary Louise Greene, *Among School Gardens* (New York: Charities Publication Committee, 1910); A. Hyatt (Alpheus Hyatt) Verrill, *Harper's Book for Young Gardeners: How to Make the Best Use of a Little Land* (New York: Harper & Brothers, 1914); and Kary Cadmus Davis, *School and Home Gardening: A Text-Book for Young People* (Philadelphia: J. B. Lippincott Company, 1918).

Genetics

Genetics (like the topic of DNA) has numerous historical threads to consider. For a general survey, see Jennifer Raff, *Origin: A Genetic History of the Americas* (New York: Hachette Book Group, Inc., 2022); Sheldon Krimsky, *GMOs Decoded: A Skeptic's View of Genetically Modified Foods* (Cambridge: MIT Press, 2019); and Jennifer Doudna and Samuel H. Sternberg, *Crack in Creation: Gene Editing and the Unthinkable Power to Control Evolution* (New York: Houghton Mifflin Harcourt, 2018). For connections to agriculture, see Margaret E. Derry, "Genetics, Biotechnology, and Breeding: North American Shorthorn Production in the Twenty-First Century," *Agricultural History* 92, no. 1 (2018): 54–77; Rebecca Mackelprang and Peggy G. Lemaux, "Genetic Engineering and Editing of Plants: An Analysis of New and Persisting Questions," *Annual Review of Plant Biology* 71 (2020): 659–87; and Peter Pringle, *Food, Inc.: Mendel to Monsanto—Promises and Perils of the Biotech Harvest* (New York: Simon & Schuster, 2005). Also, the PBS documentary *The Gene: An Intimate History* by Ken Burns and Barak Goodman offers a useful general overview of many themes related to the history of genetics.

Indigenous Knowledge/Immaterial Cultural Heritage/ Traditional Practice

Key to interpreting science in historical places and museums includes narratives of Indigenous knowledge, immaterial cultural heritage, and traditional practice. Museums and historic sites can identify the importance of practice, the role of cross-generational information sharing—all to underscore a value-added knowledge that is socially conscious. There is a growing scholarship in this context, but some key works include Devon A. Mihesuah and Elizabeth Hoover, eds., *Indigenous Food Sovereignty in the United States: Restoring Cultural Knowledge, Protecting Environments, and Regaining Health* (Norman: University of Oklahoma Press, 2019); Robin Wall Kimmerer, *Braiding Sweetgrass: Indigenous Wisdom, Scientific Knowledge, and the Teachings of Plants* (Minneapolis: Milkweed Editions, 2015); John Vaillant, *The Golden Spruce: A True Story of Myth, Madness, and Greed* (New York: W. W. Norton, 2005); Jessica Hernadez, *Fresh Banana Leaves: Healing Indigenous Landscapes Through Indigenous Science* (Berkeley, CA: North Atlantic Books, 2022); and M. Kat Anderson, *Tending the Wild: Native American Knowledge and the Management of California's Natural Resources* (Berkeley: University of California Press, 2005).

Landmarks of Science

Scientific research occurred in specific locations. Sally Gregory Kohlstedt calls for recognizing the role museums played in creation of scientific knowledge in "Essay Review: Museums: Revisiting Sites in the History of the Natural Sciences," *Journal of the History of Biology* 28, no. 1 (Spring 1995): 151–66. Scientific organizations often designate landmarks, e.g., National Historic Chemical Landmarks, administered by the American Chemical Society, https://www.acs.org/content/acs/en/education/whatischemistry/landmarks.html; National Natural Landmarks that contain "outstanding biological and geological resources," administered by the National Park Service, https://www.nps.gov/subjects/nnlandmarks/index.htm; Historic Physics Sites, administered by the American Physical Society, https://www.aps .org/programs/honors/history/historicsites/index.cfm; and Milestones in Microbiology sites, administered by the American Society for Microbiology, https://lib.guides.umbc .edu/c.php?g=836720&p=6543681.

Nature Study

For an introduction to the history of nature study, see Sally Gregory Kohlstedt, *Teaching Children Science: Hands-On Nature Study in North America, 1890–1930* (Chicago: University of Chicago Press, 2010). The majority of published primary sources readily available convey the approach taken at Cornell University, e.g., the digital archive available at https://rmc .library.cornell.edu/bailey/naturestudy/naturestudy_2.html; *Cornell Nature-Study Leaflets Being a selection, with revision, from the teachers' leaflets, home nature-study lessons, junior naturalist monthlies and other publications from the College of Agriculture, Cornell University, Ithaca, N.Y., 1896–1904*, Nature Study Bulletin No. 1 (Albany, NY: J. B. Lyon Company, Printers, 1904), available at https://www.gutenberg.org/files/43200/43200-h/43200-h.htm; and Anna Botsford Comstock compiled these into her *Handbook of Nature Study*, first ed. (1911), rev. ed. (1939) and reprinted with a foreword by Verne N. Rockcastle (Ithaca, NY: Cornell University Press, 1986).

Organic Farming

The history of organic farming includes a host of connections and interconnections for interpreting science. It often combines many of the former categories already described that connect agricultural science, environmental science, and food production and politics. For good general accounts, see Andrew N. Case, *The Organic Profit: Rodale and the Making of Marketplace Environmentalism* (Seattle: University of Washington Press, 2018); David D. Vail, "A Counterculture Agriculture: Organic Farming in a Commercial Food Age," in *A Companion to American Agricultural History*, R. Douglas Hurt, ed. (Hoboken, NJ: Wiley Blackwell, 2022), 188–99; Robin O'Sullivan, *American Organic: A Cultural History of Farming, Gardening, Shopping, and Eating* (Lawrence: University Press of Kansas, 2015); and Stephen Heyman, *The Planter of Modern Life: How an Ohio Farm Boy Conquered Literary Paris, Fed the Lost Generation, and Sowed the Seeds of the Organic Food Movement* (New York: W. W. Norton, 2020).

Pesticides, Toxicology, and Rachel Carson

The dynamic histories of pesticides, toxicology, and the scientists involved in their creation and critiques offer much for those wanting to make interpretive connections between politics, policies, environmental science, ecology, agricultural science, and domestic life. Accessible, general works include Frederick Rowe Davis, *Banned: A History of Pesticides and the Science of Toxicology* (New Haven, CT: Yale University Press, 2014); David Kinkela, *DDT and the American Century: Global Health, Environmental Politics, and the Pesticide That Changed the World* (Chapel Hill: University of North Carolina Press, 2011); Thomas Dunlap, *DDT, Silent Spring, and the Rise of Environmentalism* (Seattle: University of Washington Press, 2008); Michelle Mart, *Pesticides, A Love Story: America's Enduring Embrace of Dangerous Chemicals* (Lawrence: University Press of Kansas, 2015); and Nancy Langston, *Toxic*

Bodies: Hormone Disruptors and the Legacy of DES (New Haven, CT: Yale University Press, 2010).

On regional studies that consider these relationships, see Pete Daniel, *Toxic Drift: Pesticides and Health in the Post-World-War-II South* (Baton Rouge: Louisiana State University Press, 2005); Adam Romero, *Economic Poisoning: Industrial Waste and the Chemicalization of American Agriculture* (Oakland: University of California Press, 2021); and David D. Vail, *Chemical Lands: Pesticides, Aerial Spraying, and Health in North America's Grasslands since 1945* (Tuscaloosa: University of Alabama Press, 2018).

Rachel Carson's 1962 book *Silent Spring* spotlighted unregulated use of insecticides, especially DDT, assessed the rationale behind the design and use of toxic chemicals and the industries involved in pushing ubiquitous use, and stressed the negative consequences (long-term ecological effects to be sure) of use. On Carson, *Silent Spring*, see Robert Gottlieb, *Forcing the Spring: The Transformation of the American Environmental Movement* (Washington, DC: Island Press, 1993); Linda Lear, *Rachel Carson: Witness for Nature* (New York: Henry Holt and Company, 1997); and William Souder, *On a Farther Shore: The Life and Legacy of Rachel Carson* (New York: Crown Publishers, 2014).

Physics

To get at some of key advancements, individuals, and discoveries in physics, see David C. Cassidy, *A Short History of Physics in the American Century* (Cambridge, MA: Harvard University Press, 2013). Other works such as Daniel J. Kevles, *The Physicists: The History of a Scientific Community in Modern America* (New York: Vintage Books, 1971: repr. with a new essay on the Superconducting Super Collider, Cambridge, MA: Harvard University Press, 1995); and Roger G. Newton, *From Clockwork to Crapshoot: A History of Physics* (Cambridge, MA: Harvard University Press, 2010) make complicated theories and the evolution of the field itself (from the ancient world to the modern era) accessible for curators and the general reader alike. See also Oliver Darrigol, *From C-Numbers to Q-Numbers: The Classical Analogy in the History of Quantum Theory* (Berkeley: University of California Press, 1992); and David Kaiser, *How the Hippies Saved Physics: Science, Counterculture, and the Quantum Revival* (New York: Norton, 2011).

Public Health

Public health topics include works of local communities, rural/urban state-level histories, regional relationships, and national policies. For good general studies that begin to make these historical connections, see Mazyck Porcher Ravenel, ed., *A Half Century of Public Health* (New York: Arno Press, 1970); Michelle Kiechle, *Smell Detectives: An Olfactory History of Nineteenth-Century Urban America* (Seattle: University of Washington Press, 2017); Jennifer Lisa Koslow, *Exhibiting Health: Public Health Displays in the Progressive Era* (New Brunswick, NJ: Rutgers University Press, 2020); and Tracy Kidder, *Mountains Beyond*

Mountains: The Quest of Dr. Paul Farmer, a Man Who Would Cure the World (New York: Random House, 2009).

Scientists' Autobiographies

In addition to the historiography of scientists and their work, autobiographies and essays by scientists can reveal how and why they entered their fields of research. The best of these accounts give insight into how each scientist developed and tested ideas and the ramification they have in the public arena. This is not an exhaustive list, but a selection of scientific "armchair travel": Sydney Brenner, *Loose Ends from Current Biology* (London: Current Biology, 1997); Peter B. Medawar, *Advice to a Young Scientist* (New York: Harper & Row, 1979); Rita Levi-Montalcini, *In Praise of Imperfection: My Life and Work* (New York: Basic Books, 1988); Salvador E. Luria, *A Slot Machine, a Broken Test Tube: An Autobiography* (New York: Harper & Row, 1984); Luis W. Alvarez, *Alvarez: Adventures of a Physicist* (New York: Basic Books, 1989); Carl Djerassi, *The Pill, Pygmy Chimps, And Degas' Horse: The Remarkable Autobiography of the Award Winning Scientist Who Synthesized the Birth Control Pill* (New York: Basic Books, 1993); and Suzanne Simard, *Finding the Mother Tree: Discovering the Wisdom of the Forest* (New York: Knopf, 2021).

Autobiographies also stress challenges based on situations or identity, i.e., Rita Colwell and Sharon Bertsch McGrayne, *A Lab of One's Own: One Woman's Personal Journey Through Sexism in Science* (New York: Simon & Schuster, 2020); and Ben Barres, *The Autobiography of a Transgender Scientist* (Cambridge, MA: MIT Press, 2018).

Space Exploration

There is a plethora of histories when it comes to space exploration (consult the entries for "Women in Science" for more). A few offer accessible scholarly overviews and participants' perspectives to help lay a foundation for interpretation. See Roger D. Launius, *The Smithsonian History of Space Exploration: From the Ancient World to the Extraterrestrial Future* (Washington, DC: Smithsonian Books, 2018); and Neil M. Maher, *Apollo in the Age of Aquarius* (Cambridge, MA: Harvard University Press, 2017). And for one of the best first-person astronaut accounts of the Apollo program, see Michael Collins, *Carrying the Fire: An Astronaut's Journeys* (New York: Farrar, Straus and Giroux, 1974).

Viruses/Virology

Histories of influenza and other viruses have certainly become more in vogue with readers and visitors hoping to better understand the COVID-19 pandemic. Good general works to consult include John M. Barry, *The Great Influenza: The Story of the Deadliest Pandemic in History* (New York: Penguin, 2018); Gina Kolata, *Flu: The Story of the Great Influenza Pandemic of 1918 and the Search for the Virus That Caused It* (New York: Farrar, Straus &

Giroux, 1999); John M. Barry, *The Great Influenza: The Story of the Deadliest Pandemic in History* (New York: Penguin, 2005); David M. Oshinsky, *Polio: An American Story* (New York: Oxford University Press, 2006); Jane S. Smith, *Patenting the Sun: Polio and the Salk Vaccine: The Dramatic Story Behind One of the Greatest Achievements of Modern Science* (New York: William Morrow and Company, 1990); Angela N. H. Creager, *The Life of a Virus: Tobacco Mosaic Virus as an Experimental Model, 1930–1965* (Chicago: University of Chicago Press, 2013); Paul A. Offit, *Vaccinated: One Man's Quest to Defeat the World's Deadliest Diseases* (New York: Harper Perennial, 2008); and David Quammen, *Spillover: Animal Infections and the Next Human Pandemic* (New York: W. W. Norton, 2013). On cancer and virology, see Gregory J. Morgan, *Cancer Virus Hunters: A History of Tumor Virology* (Baltimore: Johns Hopkins University Press, 2022).

A key resource for newspaper reports throughout the United States can be found through "The American Influenza Epidemic of 1918–1919: A Digital Encyclopedia" (University of Michigan Center for the History of Medicine and Michigan Publishing, University of Michigan Library, https://www.influenzaarchive.org/). Important historical analysis on the 1918 influenza pandemic and COVID-19 pandemic can be found in Christopher McKnight Nichols et al., "Reconsidering the 1918–19 Influenza Pandemic in the Age of COVID-19," *The Journal of the Gilded Age and Progressive Era* 19 (2020): 1–31.

Water, Wetland Destruction, Drought, and Irrigation Science

Numerous themes emerge in the histories of water, drought, and irrigation science that grapple with topics around science, climate change, natural disasters, and drought. As these conditions influence visitors' everyday realities, interest should increase about how science addresses the need and how people in the past addressed similar challenges. For water purity in general, see Bhawani Venkataraman, *The Paradox of Water: The Science and Policy of Safe Drinking Water* (Berkeley: University of California Press, 2023), and for public engagement relative to the Great Lakes, see John H. Hartig, *Great Lakes Champions: Grassroots Efforts to Clean Up Polluted Watersheds* (East Lansing: Michigan State University Press, 2022). For wetland destruction, see a transnational assessment by Annie Proulx, *Fen, Bog and Swamp: A Short History of Peatland Destruction and Its Role in the Climate Crisis* (New York: Scribner, 2022).

Regarding water supply, see Stanley G. Crawford, *Mayordomo: Chronicle of an Acequia in Northern New Mexico* (Albuquerque: University of New Mexico Press, 1988); and Robert R. Crifasi, *A Land Made from Water: Appropriation and the Evolution of Colorado's Landscape, Ditches, and Water Institutions* (Boulder: University Press of Colorado, 2015). For irrigation and food production, see Mark Fiege, *Irrigated Eden: The Making of an Agricultural Landscape in the American West* (Seattle: University of Washington Press, 1999); John F. Freeman, *Adapting to the Land: A History of Agriculture in Colorado* (Louisville: University Press of Colorado, 2022); and James Earl Sherow, *Watering the Valley: Development along the High Plains Arkansas River, 1870–1950* (Lawrence: University Press of Kansas, 1990).

Weather

Local weather data can be used to document weather patterns over time, including recurring events and extremes. For background information, see Mark Monmonier, *Air Apparent: How Meteorologists Learned to Map, Predict, and Dramatize Weather* (Chicago: University of Chicago Press, 1999); James Rodger Fleming, *Meteorology in America, 1800–1870* (Baltimore: Johns Hopkins University Press, 2000); Lourdes B. Avilés, *Taken by Storm, 1938: A Social and Meteorological History of the Great New England Hurricane* (Boston: American Meteorological Society, 2013); and William B. Meyer, *Americans and Their Weather* (New York: Oxford University Press, 2000).

Women in Science

Historian of science Margaret W. Rossiter assessed women and their careers in biology and chemistry, but also in geology, home economics, physics, and psychology, in her path-breaking book, *Women Scientists in America: Struggles and Strategies until 1940* (Baltimore: Johns Hopkins University Press, 1982), followed by *Women Scientists in America: Before Affirmative Action, 1940–1972* (Baltimore: Johns Hopkins University Press, 1995), and *Women Scientists in America: Forging a New World Since 1972* (Baltimore: Johns Hopkins University Press, 2012). Historian of science Amy Sue Bix focused on women pursuing an education in the male-dominated applied science of engineering in *Girls Coming to Tech!: A History of American Engineering Education for Women* (Cambridge, MA: MIT Press, 2013). Quantum chemist Michelle Francl asked readers if they could identify ten women scientists in "Atomic Women," *Nature Chemistry* 10 (2018): 373–75. Oral history collections contain interviews with women scientists, including minority women, i.e., The HistoryMakers: The Digital Repository for the Black Experience, available at https://www.thehistorymakers.org/; and Oral History Interviews, Niels Bohr Library & Archives, American Institute of Physics, College Park, Maryland, available at https://www.aip.org/history-programs/niels-bohr-library/oral-histories.

Others have added to our understanding by documenting the roles of women in various scientific fields, recounting personal experiences, or through biographical analysis, i.e., Nina Byers and Gary Williams, *Out of the Shadows: Contributions of Twentieth Century Women to Physics* (New York: Cambridge University Press, 2006); Evelyn Fox Keller, *A Feeling for the Organism: The Life and Work of Barbara McClintock* (New York: Henry Holt, 1983); Julie Des Jardins, *The Madame Curie Complex: The Hidden History of Women in Science* (New York: The Feminist Press, 2010); Hope Jahren, *Lab Girl* (New York: Vintage Books, 2016); Walter Isaacson, *The Code Breaker: Jennifer Doudna, Gene Editing, and the Future of the Human Race* (New York: Simon & Schuster, 2021); and Cassandra Leah Quave, *Plant Hunter: A Scientist's Quest for Nature's Next Medicines* (New York: Penguin, 2021).

For the role of women in science during wartime, see Denise Kiernan, *The Girls of Atomic City: The Untold Story of the Women Who Helped Win World War II* (New York: Touchstone, 2013); Kate Moore, *The Radium Girls: The Dark Story of America's Shining Women*

(Naperville, IL: Sourcebooks, 2017); and Liza Mundy, *Code Girls: The Untold Story of the American Women Code Breakers of World War II* (New York: Hachette Book Group, 2017).

On the women who assisted in the science of space exploration, see Margot Lee Shetterly, *Hidden Figures: The American Dream and the Untold Story of the Black Women Mathematicians Who Helped Win the Space Race* (New York: Harper Collins, 2016); and Nathalia Holt, *Rise of the Rocket Girls: The Women Who Propelled Us, from Missiles to the Moon to Mars* (New York: Back Bay Books, 2017). Women scientists have written autobiographies and have been the subject of biographies. See "Scientists' Autobiographies."

Timeline of Scientific Advances, Controversies, Policies, and Legislation

This timeline lists selected scientific theories, research, breakthroughs, and legislation, in chronological order from oldest to more recent. For a comparable timeline focused on the environment, see Debra A. Reid and David D. Vail, *Interpreting the Environment at Museums and Historic Sites* (Lanham, MD: Rowman & Littlefield, 2019).

1603: Girolamo Fabrizio's work on veins.

1610: Lowest level of carbon dioxide (CO_2) in the Common Era.

1610: Galileo Galilei published *The Starry Messenger*.

1621: Robert Burton's *The Anatomy of Melancholy*.

1628: William Harvey's *On the Motion of the Heart and Blood in Animals*.

1637: René Descartes's "Illustration of an Eye" in *Discourse on Method* is published.

1660: Robert Boyle's *New Experiments Physico-Mechanical Touching the Spring of the Air*.

1662: John Graunt's *Natural and Political Observations Made Upon the Bills of Mortality*.

1665: Great plague in London.

1666: Thomas Sydenham writes on fever treatments.

1666: The Academiè des Sciences is founded in Paris.

1687: Sir Isaac Newton published *Philosophiae naturalis principia mathematica* (*Mathematical Principles of Natural Philosophy*) or the *Principia Mathematica*.

1704: Sir Isaac Newton writes *Opticks* (his most famous edition, the thirty-first query, second edition was published in **1718**).

1705: Raymond Vieussens details the internal structures of the heart, specifically the left ventricle of heart and course of coronary blood vessels.

1708: Herman Boerhaave's *Institutiones Medicae.*

1714: Gabriel David Fahrenheit designs the mercury thermometer.

1717: Giovanni Maria Lancisi proposes that mosquitoes transmit malaria.

1717: Lady Mary Wortley Montagu introduces the Turkish practice of smallpox inoculation to England.

1740: Maximum of the Little Ice Age.

1747: Albrecht von Haller's *Primae Lineae Physiologiae* (first textbook on physiology).

1747: James Lind discovers that citrus fruits can cure scurvy. His *Treatise of the Scurvy* was published in **1753**.

1752: William Smellie's *Theory and Practice or Treatise on Midwifery.*

1775: Percivall Pott identified environmental factors as a cause of cancer through his work with soot's correlation to chimney sweeps' testicular cancer.

1776: Matthew Dobson illustrates the role of sweetness in urine as an indicator of diabetes.

1780: Luigi Galvani conducts experiments with muscles and electricity.

1795: Thomas Beddoes and Humphry Davy conduct experiments with nitrous oxide (laughing gas).

1796: Edward Jenner's smallpox vaccine.

1798: Thomas Malthus's *Essay on the Principles of Population.*

1800: The Industrial Revolution.

1800s: Portable laboratories or chemistry kits (German chemicals, English construction).

1803–1806: Lewis and Clark "Voyage of Discovery" Expedition.

1812: Benjamin Rush's *Medical Inquiries and Observations upon the Diseases of the Mind.*

1816: René Laënnec invents the stethoscope.

1821: Charles Bell details facial paralysis and later (**1830**) identifies different types of nerves.

1828: Starting point of organic chemistry. Wöhler synthesis of ammonium cyanate (an inorganic chemical) into urea (an organic chemical).

1839: Photography established with daguerreotype and silver chloride–sensitive paper.

1841: F. G. J. Henle publishes on microscopic anatomy.

1846: The Smithsonian Institution is established (opened in 1855).

1849: Elizabeth Blackwell becomes first woman in America to receive a medical degree.

1858: First edition of *Gray's Anatomy* is published.

1858: Rudolf Virchow's *Celluarpathologie* (explores how every cell is a product of other cells).

1859: Charles Darwin's *The Origin of Species.*

1861: Louis Pasteur discovers anaerobic bacteria.

1864: George Perkins Marsh publishes *Man and Nature or, Physical Geography as Modified by Human Condition.*

1864: The International Red Cross is founded.

1865: Gregor Mendel's "Experiments in Plant Hybridization."

1870: The Weather Bureau of the United States is created within the US Department of War to document meteorological observations as well as anticipate storm systems. The Bureau moved to the USDA in **1890**, then the US Department of Commerce in **1940**, and then was renamed as the National Weather Service in **1970**.

1871: Charles Darwin's *Descent of Man.*

1876: Alexander Graham Bell patents the telephone.

1876: Robert Koch identifies anthrax bacterium (*Bacillus anthracis*).

1880: Thomas Edison patents the incandescent lamp.

1881: Louis Pasteur develops a vaccine for anthrax.

1882: Robert Koch isolates the tuberculosis bacterium (*Mycobacterium tuberculosis*).

1883: The US Department of Agriculture (USDA) establishes a Veterinary Division for research and the development of eradication policies for contagious animal diseases.

1884: The USDA Bureau of Animal Industry replaces the Veterinary Division.

1884: Francis Trudeau isolates the bacterium associated with tuberculosis, the first American to confirm Koch's studies.

1885: Louis Pasteur develops a vaccine for rabies virus.

1889: US Department of Agriculture (USDA) establishes an Irrigation Investigations Unit.

1889: Patrick Manson's *Tropical Disease*

1889: Marie Skłodowska Curie and Pierre Curie acquire radium from pitchblende.

1892: Ellen Swallow develops the concept of oekology based on her work in environmental chemistry and industrial health. Oekology insists on studying the earth as a "household."

1895: X-rays discovered by Wilhelm Conrad Röntgen.

1896: Henri Becquerel discovers radioactivity.

1898: Marie Skłodowska Curie and Pierre Curie discover polonium, the first element discovered due to its radioactivity.

1901: Emil Adolf von Behring wins Nobel Prize in Physiology or Medicine for his work on serum therapy and its application against diphtheria, a bacterial disease.

1901: Wilhelm Conrad Röntgen wins Nobel Prize in Physics in recognition of his discovery of X-rays.

1901: Jacobus Henricus van 't Hoff receives Nobel Prize in Chemistry for his discovery of the laws of chemical dynamics and osmotic pressure.

1905: Albert Einstein posits his theory of special relativity.

1905: US Forest Service established under the US Department of Agriculture.

1906: Pure Food and Drug Act prohibits interstate commerce of adulterated foods, drinks, and drugs.

1909–1910: Haber-Bosch process industrialized to convert atmospheric nitrogen to ammonia fertilizer.

1910: Paul Ehrlich discovers Salvarsan for syphilis (an early example of modern chemotherapy).

1910: Federal Insecticide Act regulates pesticides' production and marketing of chemicals to protect consumers and applicators from fraudulent projects.

1915: Albert Einstein discovers general relativity.

1917: The USDA Bureau of Animal Industry establishes a Tuberculosis Eradication Division.

1918: Beginning of the influenza virus pandemic. Upward of one hundred million deaths worldwide.

1919: Ernest Rutherford splits atom.

1921: Marie Stopes starts a birth-control clinic in London.

1922: First hybrid corn commercially produced and marketed.

1923: Albert Calmette and Camille Guérin create BCG vaccine for tuberculosis.

1928: Alexander Fleming discovers penicillin.

1929: USDA Munsell Color Charts; the agency's official soil-research and plant-standards color systems are still in use today.

1932: Gerhard Domagk discovers the antibiotic sulfonamidochrysoidine, the first commercially available antibiotic.

1932: Tuskegee Syphilis "Study," which continues until 1972.

1933: Ernst Ruska invents prototype for electron microscope.

1938: Laszlo Biro invents the ballpoint pen.

1938: Lise Meitner, Otto Frisch, Otto Hahn, and Fritz Strassmann discover nuclear fission.

1939: Charles Goodyear perfects vulcanized rubber.

1942: Federal Insecticide, Fungicide, and Rodenticide Act (FIFRA), to regulate production, sale and use of pesticides

1944: Selman Waksman discovers the antibiotic streptomycin.

1945: Inactivated influenza vaccine licensed for civilian use in the United States.

1945: The Trinity Test (Manhattan Project): First nuclear device detonated near Socorro, New Mexico.

1945: Atomic bombing of Hiroshima, Japan, by the United States (August 6).

1945: Atomic bombing of Nagasaki, Japan, by the United States (August 9).

1947: The US Congress passes the Federal Insecticide, Fungicide, and Rodenticide Act to regulate the potential risks of chemical applications.

1947: Maria Maynard Daly becomes the first African American woman in the United States awarded a PhD in chemistry (Columbia University).

1948: United States establishes the National Institutes of Health.

1948: The United Nations establishes the World Health Organization.

1949: Aldo Leopold's *A Sand County Almanac and Sketches Here and There* is published.

1949: Soviet Union detonates their first atomic weapon test near Semipalatinsk (modern Kazakhstan).

1950: Particle physics and accelerator lab (FermiLab) established in Batavia, Illinois.

1950: Barbara McClintock conducts fundamental cytogenetics studies using maize as a model organism.

1950: National Science Foundation established.

1951: Rosalind Franklin discovers helical form of DNA.

1952: Frank Zybach granted a US patent for his "Self-Propelled Sprinkling Irrigating Apparatus," now known as a center pivot sprinkler.

1952: United States detonates the first hydrogen bomb in Enewetak Atoll test (Marshall Islands).

1953: James Watson and Francis Crick build a DNA model.

1953: Soviet Union detonates their first hydrogen bomb in a test near Semipalatinsk in modern Kazakhstan.

1954: Jonas Salk develops a killed form of polio virus for an injectable vaccine.

1957: Albert Sabin develops live polio virus vaccine.

1957: Influenza virus pandemic.

1957: Soviet Union launches Sputnik (world's first artificial satellite).

1958: United States launches its first space satellite.

1960: US Food and Drug Administration approves the first oral contraceptive ("The Pill").

1962: Rachel Carson's *Silent Spring*.

1962: First commercial space satellite launched for television transmission in the United States and Europe (Telstar1).

1963: Clean Air Act enacted to regulate emissions and set industry standards for air pollution. This act would ultimately establish the National Ambient Air Quality Standards (NAAQS) for human and environmental protection.

1964: US Surgeon General publishes report on public health risks of smoking.

1967: Lynn Margulis develops endosymbiont theory.

1969: Robert Rayford dies of a mysterious illness akin to AIDS.

1969: Union of Concerned Scientists formed at the Massachusetts Institute of Technology.

1969: National Aeronautics and Space Administration (NASA) Moon landing (*Apollo 11*).

1969: Fire on the Cuyahoga River (Ohio) leads to legislation for the creation of the Environmental Protection Agency (EPA).

1970: The first Earth Day mobilization in the United States.

1970: The United States Environmental Protection Agency (EPA) is established.

1971: The EPA restricts lead-based paints.

1972: James Lovelock proposes the Gaia Hypothesis, which argued the planet (biosphere) is likened to a self-regulating organism.

1972: EPA bans the use of the pesticide DDT in the United States.

1972: Federal Water Pollution Control Act (Clean Water Act) enacted to "restore and maintain the chemical, physical, and biological integrity of the nation's waters" (PL 92-500).

1973: First use of recombinant DNA technology to create a genetically engineered organism (a bacterial plasmid in *Escherichia coli*).

1973: Endangered Species Act issues federal protections for threatened species as well as their habits.

1975: Scientific meeting on regulation of genetic engineering ("Asilomar Conference on Recombinant DNA" in California).

1976: Epidemics of Ebola virus disease in the Sudan and the Republic of Zaire.

1978: First in vitro fertilization or "test tube" baby.

1979: Smallpox in nature eradicated by vaccination.

1980: US Supreme Court allows patenting of living organisms.

1982: FDA approves first recombinant DNA pharmaceutical, insulin.

1981: First NASA space shuttle flight.

1981: Stem cells first cultured from mouse embryos.

1981: The US Center for Disease Control (CDC) publishes an article in its *Morbidity and Mortality Weekly Report* (*MMWR*): *Pneumocystis* Pneumonia.

1982: CDC uses the term "AIDS" (Acquired Immune Deficiency Syndrome).

1983: First genetically engineered plant (tobacco).

1983: Barbara McClintock, cytogeneticist, is the first woman to receive an unshared Nobel Prize in Physiology or Medicine.

1984: AIDS becomes the leading cause of death for all Americans ages twenty-five to forty-four.

1985: Flossie Wong-Staal (Yee Ching Wong), virologist and molecular biologist, cloned HIV (human immunodeficiency virus) and genetically mapped the entire virus, leading to the first blood tests for HIV (enzyme-linked immunosorbent assay or ELISA test).

1985: Kary Mullis invents the polymerase chain reaction (PCR) method to amplify DNA.

1985–1986: The Human Genome Project begins to take shape.

1987: "Gene Gun" tested at Cornell University for plant transformation (genetic engineering).

1990: NASA and the European Space Agency launch Hubble Telescope.

1992: Public release of hypertext, launching the World Wide Web.

1994: First genetically modified crop (tomato), approved (by USDA) for release and commercial sale. Flavr Savr, Calgene, slowed ripening and postharvest decay.

1996: First mammal cloned (a sheep named Dolly).

1996: Federal ban on leaded gasoline.

1996: AIDS is no longer the leading cause of death for all Americans ages twenty-four to forty-four due to highly active antiretroviral therapy but remains the leading cause of death for African Americans in this age group.

2001: First draft of the human genome.

2000: Drosophila (fruit fly) genome completed.

2000: Arabidopsis (plant) genome completed.

2003: Human genome completed.

2004: First commercially cloned pet (a house cat named "CC," short for Copy Cat).

2008: NASA's Phoenix Mars Lander spacecraft lands on Mars.

2012: First CRISPR-Cas9 gene editing.

2015: First genetically modified animal (salmon) for human consumption approved by FDA, then commercially marketed (**2017**).

2015: International Summit on Gene Editing (National Academy of Sciences, Washington, DC).

2016: US Congress mandates labeling of genetically modified foods.

2018: First CRISPR-edited twins born in China with a germline edit to "prevent" HIV infection.

2019: SARS-CoV-2 described, determined to be the causal agent of COVID-19.

2019: First image of a black hole in the center of Messier 87 (M87) by the Event Horizon Telescope.

2021: SARS-CoV-2 vaccine. First mRNA vaccine.

2021: First all-civilian, private space flights with SpaceShip2 and SpaceX.

2022: James Webb Space Telescope is launched and transmits images to NASA.

2022: Nuclear fusion demonstrated, a first step toward sustainable electricity.

Index

Numbers in italic denotes a figure

biodiversity, 127, *129*

bioengineering. *See* genetic engineering (GE)

bioethical concerns, using CRISPR, 106, 108n9

biolistic particle delivery system, 105, 107

biomedical research underrepresentation, 90

Bio–Rad, 105, 108n7

biosecurity, 38, 41n9

biosphere, 201

bird flu. *See* avian influenza virus

bird watching festivals, 152

Bird-Lore (magazine), 163

birds: yellow–billed cuckoo, 149; educational lessons about, 163; migration, 25–26, 152; watching, 151–53; wild, 40, 41–42n11, 163. *See also* chickens

Biro, Laszlo, 199

Birth-control clinic, 198

Bishop Museum, 177

black box, understanding of technology, 80, 105, 107, 108n10

Black Death, 3

Black gay men, 80, 89–93

Black History Month, 98

black hole, 202

Black Men's Xchange, 92

Blackwell, Elizabeth, 197

boards of education, 139

boards of health, 111, 113

Boerhaave, Herman, *Institutiones Medicae*, 196

Bost, Darius, *Evidence of Being*, 91

Boundaries of Blackness (Cohen), 90

Boyer, Paul, *By the Bomb's Early Light*, 84

Boyle, Robert, 22; *New Experiments Physico–Mechanical Touching the Spring of Air*, 195

Brazil, 75

Brooklyn Institute of Arts and Sciences (now Brooklyn Museum), 164

Brooklyn Public Library, 166

Brotha's and Sista's, Inc., 92

Brother to Brother, 92

Brown, Dr. Roscoe Conkling, 60, 61

Browne, Malcomb W., 97

buildings, xv, 16, 44, 169; built environment, 175n9; houses, 16; hospital buildings, 31; medical facilities, 31, 54

Bulletin of the Atomic Scientists Doomsday Clock, xx, xxivn12

Burbank, Luther, 104

Burpee, W. Atlee, 104

Burton, Robert, *The Anatomy of Melancholy*, 195

business history, 98

butterflies, 152; collecting, 164; net, *164*; swallowtail, 149

By the Bomb's Early Light (Boyer), 84

Cache la Poudre River, 67; National Heritage Area, 69

Cajete, Gregory, 161

Calmette, Albert, 198

Calgene, 202

California, xix, 105, 135, 201; Berkeley, 138; Los Angeles, 23–24

Canada: Northwest Territories, 127

canals, 65, 67, 68, 174

cancer: cells harvested without permission, 127; testicular, 196; treatments, 171

carbon dioxide, 4; cartridges, 105; lowest levels of, 3, 195

Carpenter, Ford, *The Land of the Beckoning Climate*, 24

Carpenter, Louis, 67

Carruthers, George R., 96

Carson, Rachel, 166; *Silent Spring*, 200; suggested reading, 189–90

Carver, George Washington, 24, 96, 136, 165; *Nature Study and Children's Gardens*, 136, 165; *Nature Study for Gardening for Rural Schools*, 136; *Nature's Garden for Victory and Peace*, 165; *Need for Scientific Agriculture in the South*, 165; *Progressive Nature Study*, 136

case studies, xviii, 123, 135

CDC. *See* Centers for Disease Control and Prevention (CDC)

colonization: colonial expansion, 112; colonialism, 23; disruption of knowledge as result of, 162; Euro–colonial science, damage of, 127

color line, 98

Colorado: 66, 67, 68, 69, 172; Bellvue, 67, 69; Fort Collins, 65, 67

Colorado Agricultural College (CAC), 65, 67–68. *See also* Colorado State University (CSU)

Colorado College, xi

Colorado River Basin, 68

Colorado State University (CSU), 65, 69; Agricultural and Natural Resources Archive, 69; Experiment Station, 69; Extension, 69; University Archives, 69; Water Resources Archive, 67, 69, 70. *See also* Colorado Agricultural College (CAC)

Colorado–Big Thompson Project, 68

Columbia University, 199

Committee of Black Gay Men, 91

commodities: bartering, 72; enhanced of, 72, 75; genetically engineered, 105; global trade of, 112; newly developed, 72; oversupply of, 71; production of, 71, 133; sales of, 72–73

community/communities, xi, 85, 173; African American, 35, 58, 60–62, 90, 172, 175n6; building, 10, 91, 136; clean–up, 57, 58; conversations with, xv, xxi; disease avoidance in, 50, 53; diseased, 32; engagement of, xxii–xxiiin2, 2, 12, 80, 140, 177; farming, 109–10, 113; fishing, 161; gardening and gardens, 138, 140; global, xii; health of, 32, 38, 175n9; medical, 50; norms of, 112; nurturing, 91; planning, 90; science affecting, xvii; science project for, 24, 26; survival of, 18, 91; wartime, 81, 83

Comstock Publishing Company, 163

Comstock, Anna Botsford, 163; *Handbook of Nature–Study for Teachers and Parents*, 163; *How to Know the Butterflies*, 163; *Insect Life*, 163; *Manual for the Study of Insects*, 163

Comstock, John Henry, 163; *How to Know the Butterflies*, 163; *Insect Life*, 163; *Manual for the Study of Insects*, 163

conservation: health, 61; landscapes, 43; movement, xii; nature, xii, 43, 144, 163; wilderness, 163;

consumer: protection, 38, 146, 198; rights, 175n9; technology, 84

consumption. *See* tuberculosis (*Mycobacterium tuberculosis*)

Continental Divide, 68

Conversation/conversations, 12. *See also* audience, community/communities; interpretive aids, source materials

Conway, Erik, *Merchants of Doubt*, xxi

Cook County Normal School, 135, 162

Cooper, Mario, 92

corn (*Zea mays*; maize), 44, 105, *106*, 138, 199; genetically engineered, 105; hybrid, 104, 198; introduction, 43, 46; science topics relative to, *106*.

Cornell University, 105, 135, 140n2, 163, 202; New York State College of Agriculture, 135

corporations, 12, 75, 104, 113, 163, 171

cotton: boll weevil, 165; crop, 136; cottonseed meal, 39; genetically engineered, 105; gin, 153

counterculture, 144

courthouses, county, 30

COVID–19, 9, 93, 174, 176n11, 202

Cowan, Ruth Schwartz, *More Work for Mother*, 85, 120, 121n1

craft-based society, 110–11

Crawford, Stanley G., 176n11

Creighton University, 30

Crenshaw, Kimberlé Williams, 175n7

Crick, Francis, 200

CRISPR–Cas9 gene editing, 108n9, 202

crops, 112; become weeds, 46; cash, 138; clover, 46; cultivation of, 134; on furrows and ridges, 17–18; genetic engineering of, 104, 107; genetically engineered, 104–7, 202; irrigation, 65–66, 70; listed as bioengineered, 105; native, 44; rotation of,

docent, 82, 169

do–it–yourself (DIY): chemistry, 97; gardens, 144; gene gun, 108n11

Domagk, Gerhard, 199

domestic economists. *See* scientific vocations, home economists

domestic economy, 111

domestic life, 121n1, 184

domestic science, 111, 114–15n5; suggested reading, 184–85

dormitories, 31, 51

double helix, 103, *145*. *See also* DNA

downforce, 156

drag, 156

drawing, relationship–building with, 125–27

drought, 6, 18, 36, 69, 70, 146, 174; suggested reading, 192; tolerant, 104

drugs: antiretroviral, 90, 202; insulin, 106, 201; legislation and policy, xv, 113, 198, 200; prophylactic, 90; tuberculosis treatment, 53–54; users of, 80, 90. *See also* tuberculosis

DuPont, 108n7. *See also* E. I. du Pont de Nemours, 105

Dust Bowl, 69

E. I. du Pont de Nemours, 105

East Asian plant introduction, 43

eBird, 152–53

ecoliteracy, 139

ecology, 70, 103, 111, 127, 149, 172; disruption of, 46; knowledge of, 161–62, 166; restoration of, 152–53

economic: activity, 22; barriers, 13n10, 58, 172; crisis, 134; development, 73–74, 136, 162; distinctions, 139; forces, 173; ideology, 72; markets, 72, 75; privilege, 13n10; profit, 39, 146; status, 172

economics, xix, xxiin1, 11, 36, 144

ecosystem, 47n9, 161

Edible Education, 138

Edible Schoolyard Project, 138

The Edison Institute, 138. *See also* The Henry Ford

Edison, Thomas, 197

education: convergence, xv, xxiin1, 124; edible, 123, 138, 140; environmental, 139–140, 166; formal, xvi, xx, 11, 99, 139, 149, 161–62, 165; high school, 135; informal, xv, xvi, 11, 161; lesson plans, 156, 107n3; outdoor, 35, 126, 134–35, 138–39; private, 30, 98, 135; public, 12, 98, 175n9

educational methods and theories, 134

educational reforms, 134, 139

educational toys, 79, 95–99, 100n3

Effland, Anne, 145

eggbeater, 80, 117, *118*, 119–20; Ekco stainless steel eggbeater, *118*; Monroe eggbeater, 117, *118*, 119. *See also* eggs; electric mixer

eggs: beating, 80, 117; prices, 41–42n11; production, 37–39, 42n12; transmitting disease, 41n9; whites whipped, 117, 120

Egyptian mummies, xii

Ehrenreich, Barbara, *For Her Own Good*, 84–85

Ehrlich, Paul, 198

Einstein, Albert, 198

Eisenhower, Dwight D., 72

electric mixer, *118*, 119–20; "Magic Maid" electric mixer, *118*

electricity, 196, 202; suggested reading, 185

electrification, 120

ELISA test. *See* enzyme–linked immunosorbent assay (ELISA) test

emission regulations, 200

empathy, 7, 83, 149, 161

empowerment, self and communal, 91

endangered species, 67. *See also* threatened species

Endangered Species Act, 201

Endfield, Georgina, 25

Enewetak Atoll, Marshall Islands, 200

England, 196

English, Deirdre, *For Her Own Good*, 84–85

environmental: change, 11, 25, 161; chemistry, 198; conditions, 5; consequences, 107, 196; contamination, xix; context, 6–7, 29; degradation, 166; educators and education,

139–140, 166; exploitation, 144; health, 38; historian, 25; influences, 49; movements, xviii; practices, 53; protection, xii, 166, 200

Environmental Protection Agency (EPA), 175n9, 176n10, 200

environmental science, 38; suggested reading, 185

environmental studies, xx

environmental topics, 143, 144

environmentalism, 169

enzyme–linked immunosorbent assay (ELISA) test, 201

EPA. *See* Environmental Protection Agency (EPA)

epidemic: Ebola, 201; HIV/AIDS, 80, 89, 92, 93. *See also* diseases, HIV/AIDS; pandemic; tuberculosis

erosion, 146

Escherichia coli, 201

Essay on the Principles of Population (Malthus), 196

European Space Agency, 202

Event Horizon Telescope, 202

Evidence of Being (Bost), 91

Ewing, Oscar R., 61

exhibitions: *Becoming Jane: The Evolution of Dr. Jane Goodall*, xii; *Becoming Weatherwise*, 23, 24; changing based on community interaction, xxii–xxiiin2; climate change, xii, xxii–xxiiin2, 4–7; digital, 31; *Driven to Win: Racing in America*, xxi, 123–24, 155–56, 159n5; *Household Insects*, 164; interactive, xii, 126; Julia Child's kitchen, 121n3; *Lenaphoking*, 166; linking science and history, *xvi*; museum, 126, 149, 151, 173; online, 6; planning, xvii; pop–up, 99; *Science at Play*, 98; science–based, 169, 173; traveling, xi–xii, 23; *Treasures of NOAA's Ark*, 23; virtual, 98, 184; "Weathered History" project, 6; *The White Plague Comes to Kearney: A History of the Nebraska State Hospital for Tuberculosis*, 31–32. *See also* insects

experiment station, 69, 144, 145

experimental archaeology, xxi, 15–16, 18

experimentation, human, 171. *See also* Tuskegee Normal and Industrial Institute

exploitation, environmental, 144

Fabrizio, Girolamo, 195

factory production, 112

Fahrenheit, Gabriel David, 196

fairs: county, 37, 138; fancy, 37; science, 97; state, 138

Fairview Garden School Association (Yonkers, New York), 137

famine, 6, 72

FAO Schwarz, 98

farm families. *See* family farms

"farm to fork," 111

farmers, 44, 52, 133, 145, 146; of color disparaged, 144; irrigation by, 66–67, 69; plant history and, 43; tips for, 39; as trusted source of food, 109, 110; urban, 39

Farmers Bulletin, 39

farming, 174; community, 111, 113, 135, 147n3; environmental changes effecting, 144; practices, 1, 17–18, 135, 146; restricted, 136; transformation, 109

farms, xi, 37, 39; diseases at, 144; experimental, 104, 165; family, 17, 113, 134, 135, 136, 143, 165; living history, 44, 133; loss of, 134; mechanized equipment on, 112; organic, 144; practices at, 135; processes used on, 33n1; reconstruction of, 43–44

FDA. *See* US Food and Drug Administration (FDA)

Federal Insecticide Act, 198

Federal Insecticide, Fungicide, and Rodenticide Act, 199

Federal Water Pollution Control Act (Clean Water Act), 201

feeding programs, 72, 75, 144

FermiLab, 199

Ferriere, Madeleine, 110

fertilizers, 72, 198; organic, 17, 165

fever treatments, 195

Global South, 175n3

global warming, 4, 7. *See also* climate change

Glover, Robert, 69

Godey's, 113

Goldin, Dan, 175n7

gonorrhea, 59

Good Housekeeping, 113

Good Neighbors, Inc., 60

Goodall, Jane, xii

Goodyear, Charles, 199

Government. *See* funding, government policies, US departments

government policies, 36, 71–75, 144, 153, 162

grants, 68

grasslands, 174

Graunt, John, *Natural and Political Observations Made Upon the Bills of Mortality*, 195

Gray's Anatomy, 197

Great plague, 195

Great Plains, 31

Great Society reforms, 72

Green Revolution, 104, 144, 147n3

Greenfield Village, 138, *139*, 192. *See also* The Edison Institute; The Henry Ford

Greenhoe, James, 99–100n3

greenhouse gas emissions, 4, 106

Gross, Magdalena, 82–83

groundwater, 69

growers, commercial, 133

Guérin, Camille, 198

Haber-Bosch, 198

Habermas, Jurgen, 12

Hahn, Otto, 199

Haines, Ann, 24

hair care, 96

Hampton Normal and Agricultural Institute, 136

Handbook of Nature-Study for Teachers and Parents (Comstock), 163

"hands-on" education, 97, 124, 163, 173; instruction, 138

"happy thought," xvi, xxiiin3

Harlem Hospital, 60

Harper's Book for Young Gardeners (Verrill), 137–38

Harris, Craig G., 91

Harvard College, 135

Harvard University, Global Health Education and Learning Incubator, 176n10

harvest, 17, 18, 44, 138, 153, 202; records, 5

Harvey, William, *On the Motion of the Heart and Blood of Animals*, 195

Hatch Act, 135

Hawai'i: Honolulu, 177

Head, Kenneth, 99–100n3

health, xii, xviii, xix, 29, 30, 36, 79, 146; advocacy, 62; African American, 57–58, 61–62, 89–91, 172; animal, 38–40, 106, 152; boards of, 111, 113; clinics, 58; community, 175n9; demonstrations, 58; departments, 57–58; education, 38–40, 58, 60–61; effects of consuming GE food, 105–6; equity, 91; examinations for free, 61; forest, 45; history, 32; improvements, 90, 96, 134–35; industrial, 198; officials, 61; pageant, 60; plant, 43; policies, 89; poor, 51, 74; posters, 59, *59*, 60–61; public, xii, xix, xx, 1, 35, 49, 53, 90, 106, 110, 173; reform, 62, 135; regional economies of, 23, 51–53; scalp, 96; soil, 146; women's, 30, 85. *See also* National Negro Health Week; One Health

health care, 175n9; activism, 61–62; cost, 40; efforts increased, 58–59; HIV/AIDS, 80; inadequate, xxi, 35, 58, 62; rural, 31; unavailable, 53, 58; unequal, 35, 61, 80, 89–90

"Health Week" poster, 59, *59*

heart, 196

heat management, 156–57

heirloom and heritage varieties, 45, 133

HeLa cells, 30, 127

heliocentric model, 127

Helios Gene Gun, 105, 108n7

Hemphill, Essex, *Ceremonies*, 91

Henle, F. G. J., 196

working outside of, 80, 136, 140. *See also* Manhattan Project; Cold Spring Harbor Laboratory

labor–saving devices, 117, 120, 121n1

Lacks, Henrietta, 30, 127

Ladies' Home Journal, 113

Laënnec, René, 196

Lancisi, Giovanni Maria, 196

land grant colleges, 65, 67, 135, 136, 163, 180, 184; expertise, 73–74, 144

land management, xviii, 172

The Land of Sunshine, 51

The Land of the Beckoning Climate (Carpenter), 24

land practices, 143

landlord, 17, 18, 136

landmarks of science, suggested reading, 188

landowners, 35, 136

landscapes, xv, 1, 46; changing, 174; conservation, 43; cultural heritage, 5; reconstruction, 16, *17*, 44; restoration, 43, 165; surveys of, 17

laughing gas (nitrous oxide), 196

Lauresham Open–Air Laboratory for Experimental Archaeology, 16, *17*

Lavine, Matthew, *The First Atomic Age*, 84

lead contamination, 173–74, 176n10

lead-based paint, 200

Lee, Danielle, 138

Leopold, Aldo, *A Sand County Almanac and Sketches Here and There* (Leopold), 199

lesbians, 90, 93; African American, 92

Lewis, Meriwether, 196; portrayed, 149, 151

LGBTQ+, 91

light pollution, 9

Lincoln Log Cabin State Historic Site, Fall Harvest Fest, 151

Lincoln, Abraham, 111

Lincoln, Thomas (Abraham's father), 151

Lind, James, *Treatise of the Scurvy*, 196

Little Ice Age, 3, 196

The Little Red Hen, 37

livestock, 16, 39, 41n1; grazing, 174; pasturage, 46

local food movement, 114

locavor movement, 106

Lohr, Abby, 139

London, England, 177, 195, 198

Los Alamos Historical Society, 81, 86n9

Los Alamos History Museum, 81–85, *85*, 87n1, 87n18

Louisiana Purchase Exposition, 137

Lovelock, James, 201

Lutterlough, Sophie, *170*

Lyndon B. Johnson Presidential Library, memoranda from, 72

Mackenzie, Margaret, 24

Madame Curie (Curie), 96

Madison, James, weather observations of, 23

magic, 98–99, 101–2n13

mail-order business, 39

Maine State Library, 30

maize. *See* corn (*Zea mays*)

Making Technology Masculine (Oldenziel), 84

malnutrition, 75. *See also* nutrition

Malone, Annie Turnbo, 96

Malthus, Thomas, *Essay on the Principles of Population*, 196

Man and Nature (Marsh), 197

Manhattan Project, xix, 81–85, *85*, 199; Chemistry Division of the Metallurgical Laboratory, 99n1

"Manifest Destiny," 113

Mann, Mary Tyler Peabody (Mrs. Horace), *The Public–School Garden*, 134

manors, 18

Manson, Patrick, *Tropical Diseases*, 197

Manual for the Study of Insects (Comstock and Comstock), 163

maps, 23, 31, 69

marginalization, 62, 79–80, 89–90, 92, 172–73

Margulis, Lynn, 200

marketplaces, regulated, 114

markets. *See* stores and markets

Mars, 202

Marsh, George Perkins, *Man and Nature*, 197

Martin Luther King Jr. Middle School (Berkeley, California), 138

Martin, Martha, 165

Maryland, 60, Hagerstown, 99–100n3

mass media, 12

Massachusetts, 111; Boston, 57– 58, 111, 162; Sturbridge, 44; Walden Pond (Concord), 25

Massachusetts Institute of Technology, 200

material culture, 119, 124, 144, 146, 147n4. *See also* artifacts, buildings

Maximillian, Prince, portrayed, 149

McClintock, Barbara, 199, 201

McMurry, Charles A., 138

Mead, Elwood, 67

meadows, 16

meat: beef, 39; pork, 39

medical experiments, 30

Medical Inquiries and Observations upon the Diseases of the Mind (Rush), 196

medical intervention, 29

medical officer, *52*

medical practice, traditional, 29

medical treatment, 29, 61

medicine, 1, 103, 171, 173; alternative, 165; genetically engineered, 106; history of, 31, 84; plants used as, 44; preventative, 58; therapeutic, xx; veterinary, 103

medieval: agriculture, system of, 1, 16–18; Carolingian era, 16; furrows, 16–18; ridges, 16–18

Meitner, Lise, 199

men: African American gay, 89–91; Anglo–American gay, 90; disacknowledgment of gay, 89

men who have sex with men (MSM), 90–91

Mendel, Gregor, 197; portrayed, 149, 152

mental disability, 90

Merchants of Doubt (Oreskes and Conway), xxi

Merck Veterinary Manual, 41n9

Merian, Maria Sibylla, 127

Messier 87 (M87), 202

meteorology. *See* weather

Metropolitan Life Insurance Company, 58

Mexican, demographics in New Mexico, 52

Mexico, 35, 46

Michigan: Dearborn, 138, *139*; Flint, 173–74, 176n10

microbes, suggested reading, 183

microclimate, 16

microorganisms, 41n3, 50, 54n2

microscope: electron, 199; toy, 99–100n3

microwave, 120

Midwest, 112, 166–67n1

Midwest Open Air Museums Coordinating Council, 166–67n1

midwife and midwifery, 30, 196

A Midwife's Tale (Ulrich), 30

military hospitals, for tuberculosis patients, 50–51

milk, 39, 110; nonfat dry, 74

Miller, Harry, 157–58

Miller, Jeanine Head, 99–100n3

Miller-Tucker, Inc, 158

minerals, 74

Mississippi County Tuberculosis Association, 61

Missouri: St. Louis, 90, 137

Mitman, Greg, 23

Mock, Cary, 25

modernization theory, 72–73

Montagu, Lady Mary Wortley, 196

Montessori schools, 138

Moon landing, 200

Moon, Michelle, 71–72

moral reform, 134

More Work for Mother (Cowan), 85, 121n1

Morrill Land–Grant College Act, 135. *See* land grant colleges

mosquitoes, 196

motherhood, 85

mouse embryos, 201

moviemaking industry, 24

mRNA vaccine, 202

MSM. *See* men who have sex with men (MSM)

Muir, John, 163

Mullis, Kary, 201

organic: farming, suggested reading, 189; fertilization, 17, 165

organizations: advocacy and support, 80, 91; African American gay men, 80, 91–92; civic, 60; fraternal, 12, 60–61; health service, 92; irrigation, 69; professional, xxiiin7, 69; religious, 72; voluntary, 57, 72; women's 58, 61, 158

organoleptic metrics, 110

The Origin of Species (Darwin), 150, 197

The Origin (Stone), 150

Orion Nebula, 9

Oslo Museum of Natural History, xviii, 3

osmotic pressure, 198

OSV. *See* Old Sturbridge Village (OSV)

out–of–school experiences, 98, 136, *137*, 164

OutWrite, 93

P.L. 480. *See* Public Law 480 (P.L. 480)

Past Global Changes (PAGES), 6

PAGES–CRIAS working group, 7n1

painting, 58

paints, lead–based, 200

Pakistan, 69

pandemic, 49; COVID–19, 9, 93; influenza virus, 198, 200

paper, silver chloride–sensitive, 196

paralysis, 196

Paris, France, 135, 195

parks: as a place for learning, xii, 10–11; public service core of, 12; urban, 164

Parshall flume, *66*, 67. *See also* flume

Parshall, Ralph, 65–66, *66*, 67–70

Particle Delivery System, 108n11; particle gun, 105, *105*, 107, 108n11

particle physics and accelerator lab, 199

partisan politics, 5

Pasteur, Louis, 54n2, 197

patent: eggbeater, 80, 117, *118*; gene gun, 105, 108n5; incandescent lamp, 197; living organisms, 201; spork, 119; sprinkler, 200; telephone, 197

pathogens, 104, 145, 41–42n11; consumers affected by, 37–38; effects with tuberculosis, 50; plant, 54n2, 105; traveling, 39–40

patients, 29–31, 32, 49–51, 53

PCP. *See Pneumocystis carinii* pneumonia (PCP)

PCR. *See* polymerase chain reaction (PCR) method

Pearson, T. Gilbert, *Stories of Bird Life*, 163

peasants, 18

pen, ballpoint, 199

Pennsylvania; California, 135

penicillin, 198

"The People's Department," 111

Perfection Lighting Company, *164*

pesticides, 144, 171; banned, 201; regulation of, 198, 199; suggested reading, 189–90

pests, 44, 74, 105, 145, 164, 165

Petra, xii

Pfeiffer, Martin, 82

Pfister, Christian, 5

pharmacopeia, 30

phenotypes, 104

Philadelphia, Pennsylvania, 23, 24, 92, 173, 177

philanthropists, 135, 165

Phoenix Mars Lander, 202

photography: established, 196; vendors, 67

photomicrograph, 164

photosynthesis, enhanced, 104

phylogenetic tree, *129*

physics, suggested reading, 190

"The Pill," 200

pitchblende, 197

placemaking, 91

Plains states, 112

planets, 9

plants, 16, 143, 151, 163, 165; biotechnology used on, 104; breeders of, 104–5; brewing use of, 44; collecting of, 96, 164; color standard for, 199; cultivars, 133; drought tolerant, 104; dyeing use of, 44, 151; fiber use for, 44, 103, 151; for food, 44, 46, 123,

astrophysics, 9, 179–80; atmospheric science, 66, 70; bacteriology, 39; biochemistry, 104; biology, xx, 49, 96, 103–5, 134, 139, 144; biomedical science, xii; biotechnology, 107, 107n3; botanical analysis, 136; botany, 103, 139, 153; chemistry, xx, 66, 80, 99, 101–2n13, 104, 120, *129*, 144, 153; citizen science, 23, 124, 152–53, 166; civil engineering, 172; climate history, xxi, 5–6; climate science, 4–5, 7, 11, 23; community science, 23, 24, 26; cooking science, 120; cytogenetics, 199; earth science, xxiin1, environmental chemistry, 198; environmental science, 38; extension science, 144; food science, 121n1; forensic science, 103; geology, 172; home economics, 111; horticulture, 133; hydraulics, 66, 68; hydrology, 66, 68; invertebrate zoology, 163; irrigation science 65–67, 69–70, 192; irrigation science, suggested reading, 192; materials science, 99, 119, 156; medical science, xxi, 29–30, 31–32, 139; meteorology, 24; microbiology, 49, 80, 144; molecular biology, 104, 107, 107n3, 174, 176n11; molecular genetics, 104; molecular genetics, 104, 107; mycology, 96; natural science, 5–6, 15, 134, 139, 144, 161, 163; organic chemistry, 196; organic science, 151; ornithology, 152; physical science, 134; physics, xx, xxi, 95, 104, 120, 153, 156, 158; physiology, 196; plant pathology, 96, 144; plant science, 107, 134; robotics, 107; space exploration, 191, 194; toxicology, 189–90; veterinary medicine, 103, 197; virology, 191–92; wildlife biology, 166–67n1; zoology, 139

scientific knowledge and expertise, suggested reading, 178–79

scientific vocation inspiration, 96, 98, 99–100n3

scientific vocations: agricultural engineer, 66; agricultural engineering professors, 68; agricultural scientist, 135, 136, 143; agriculturalist, 75; agronomist, 72;

army surgeon, 23; astronaut, 175n7; bacteriologist, 50; biologist, xi, 43, 162; biology teacher, 138; botanist, 24, 46, 162; chemist, 24, 79, 95, 111–12; chemistry professor, 97; citizen scientists, 136, 152, 166; civil engineering professor, 68; climate historian, 6; community scientist, 24; conservationist, xii; consulting scientist, 173; crop science, 70; cytogeneticist, 201; dendroclimatologist, 6; ecologist, 25, 43–44, 162; engineers, 69, 80, 97, 99–100n3, 101, 156, 159, 175n7; entomologist, 163, *170*; environmental educator, 139; environmentalist, 144, 166; evolutionary biologist, 150; evolutionary geneticist, *128*; explorer, xii; extension specialist, 143; floriculturist, 45; geographer, xii, 25; health practitioner, 29–30, 41–42n11, 57; home economist, 111; horticulturalist, 104, 135; inventor, 98, 117, 119, 159; laboratory scientist, 174; mechanical engineer, 105; medical professional, 60, 135, *137*; meteorologist, 24, 25; molecular biologist, 105, 201; naturalist, 22, 149, *164*, 166n1; nuclear physicist, 101n10; nurse, 30–31, 58, 60; nursing teacher, 61; ornithologist, 152, *152*, 163; pharmacist, 30, 113; physicist, 95, 96, 101n10; professor, 65, 67, 93, 98; research scientist, 105; science educator, 128, 156; technologist, xii; theoretical chemist, 96; veterinary health practitioners, 41–42n11; virologist, 41–42n11, 201. *See also* vocations

scientist, first use of the word, xvi

seasons, 174; observation of, 24, 25–26; planting and growing, 44, 66, 104, 133; variation in, 4, 6–7

seed: bulletins, 104; catalogs, 104; companies, 104; exports, 75; hybrid, 72; maintenance, 133; mills, 46; packets, 104; production, 133; purveyor, 75

segregation, 24, 35, 61, 144;

self-discovery, 97;

self-help, 74, 136

Dozen," 60; National Negro Health Week Committee, 60; Office of Negro Health Work, 60; special programs branch, 61

US Supreme Court, 201

US Weather Bureau, 21, *22*, 23, 24, 197; Cooperative Observer Program, 23. *See also* National Weather Service

USA Today, 177

USDA. *See* US Department of Agriculture (USDA)

USPHS. *See* US Public Health Service (USPHS)

utensils, culinary, 119, 121n1, 121n5

utility company requirements, 175n9

vaccines, 196, 197, 198, 199, 200, 202

Van 't Hoff, Jacobus Henricus, 198

Vasconcelos, Angélica, 119

Vatican, 127

vegetables, 39, 133; bean, 46; collard leaves, 136; Jerusalem artichoke, 44; potatoes, 138; shelf–life, 104; sold for students, 138; squash, 44; turnips, 44; wild, 165. *See also* tomatoes

veins, 195

vendor catalogs, 67

venereal disease, 35, 59

Verrill, Alpheus Hyatt, *Harper's Book for Young Gardeners*, 137–38

Vetter, Christian, 95

videotape, educational, 68–69

Vieussens, Raymond, 196

violence, 44, 81, 82

Virchow, Rudolf, *Celluarpathologie*, 197

Virginia, 58; Richmond, 60

Virginia Union University, 60

virology, suggested reading, 191–92

viruses, 104; Ebola, 201; influenza, 198, 200; polio, 200; rabies, 197; suggested reading, 191–92; tobacco mosaic, 105. *See also* avian influenza virus (HPAIV); Human Immunodeficiency Virus (HIV); SARS–CoV–2

vitamins, 74–75

vocations: administrators, 51, 61, 73, 134, 135, 138, 175n7; arts educator, *129*; baker, 110, 153; butcher, 110, 153; camp counselor, 149; candlestick maker, 153; care provider, 89; chef, 121n3, 138; consultant, 60; crabbers, 161–62, 165; dairymen, 110; dressmaker, 153; educators, xii, 43, 97, 134–35, 138, 140n3, 149, 162, 165; elevator operator, *170*; environmental historian, 25; gardener, 43–44; graphic artist, 98; hospital cook, 31;housewife, 111, 117, 120; journalist, 90; lecturer, 60; national park ranger, 9; novelist, 93; park intern, *10*; park manager, 166; program manager, 75; public historian, 71, 75; restauranteur, 138; sailors, 50; school administrators, 162, 164; soldier, 50; spice dealers, 110; tailor, 153; tinker, 153. *See also* curators; farmers; scientific vocations; teachers

Der volksschulgarten (Schwab), 134; translated, 134

voluntary organizations, 57, 72

volunteers, 23, 24, 30, 86n1, 166–67n1

von Behring, Emil Adolf, 198

von Haller, Albrecht, *Primae Lineae Physiologiae*, 196

"Voyage of Discovery," 196

The Voyage of the Beagle (Darwin), 150

vulcanized rubber, 199

Wade House Historic Site, 166–67n1

Waksman, Selman, 199

Wallace, Henry, 143

war, 81, 134; Austro–Prussian War, 134; Franco–Prussian War, 134; nuclear, 83; Vietnam War, 73. *See also* Cold War; World War I; World War II

War on Poverty, 74

Washington, Booker T., 58, 61; founds National Negro Health Week, 57, 61

Washington, D.C., 23, 59, 60, 121n3, 177, 202

Washington: La Conner, xxii–xxiiin2

waste, 31, 39, 119

water, 66, 110; allocation, 66, 172; community report, 175n9; company, 175n9; contaminated, 41n9, 173–74, 176n10; forecasting, 68; history, 67, 69–70; legislation, 175n9, 201; measuring, 65–67, *66*, 69, 174; mills, 153; quality, 175n9; quality studies, 67, 173; rights, 35, 66, 174; sources, 35, 174; suggested reading, 192; supply, 66, 69, 174; surface, 173; systems, 173–74; tap, 173. *See also* Colorado–Big Thompson Project; groundwater; irrigation

Waters, Alice, 138

waterways, contaminated, 174

Watson, James, 200

WDBO radio, 61

Weart, Spencer, *Nuclear Fear*, 84

weather, 1, 50, 110; almanacs, 5; "American," 23; art and artifacts, 6; computer models, 25; diaries, 5, 23, 24, 25; disease related, 23; drought, 6, 18, 36, 69, 70, 146, 174; maps, 23; national services formed, 23; patterns, 17; rainfall, 24, 65; station, 24; suggested reading, 193; data: hindcasting, 25; measuring, 21, 23, 25; observer and observations, xv, 1, 21, 22, 23, 24–25, 26, 197; reanalyses of, 25; reconstruction, 6, 7, 25; records use, 21–26; extreme: 5, 18, 25; hurricanes, 25; tropical cyclones and storms, 25; storm prediction, 197; instruments: xxi, 21, 23; barometers, 22–23; rain gauge, 21, *22;* thermometers: 21, *22*, 22, 23; mercury in, 196. *See also* climate, climate change, exhibitions, reconstruction

weeds. *See* flowers and weeds

WEIRD (western, educated, industrialized, rich and democratic), 169–71, 175n3

Wellcome Collection, 177

wellness, 29, 30, 91, 146; networks, 89

wells, 69, 173

Western Illinois University, 166–67n1

wetland destruction, suggested reading, 192

wheat, 75, 105, 119, 144

Whewell, William, xvi, xxiiin3

Whiggish history. *See* presentism

Whirlpool Galaxy, 9

whisk, 80, 117, 119–20

White House Council on Environmental Quality, 162

White House Office for Food for Peace, 72, 75

White House Office of Science and Technology Policy, 162

The White Plague Comes to Kearney: A History of the Nebraska State Hospital for Tuberculosis, 31–32

White Plague, 135. *See also* tuberculosis

white supremacy, 136

white. *See* Anglo–American

whiteness, presumed, 44

whitewashing science and art, 127

wildfires, xviii

Wiley, Harvey Washington, *112*, 145

Williamson, Fiona Clare, 25

Wilson, Bee, 119–20, 121n4

Wilson, Phil, 92

Wilson, Woodrow, 24

wind erosion, 146

Woman's National Farm and Garden Association, 138; Lancaster unit, *38*

women: clubs and organizations, 58, 61, 138; disacknowledgment of, 79, 84, 98; gendered roles of, 144; health of, 30, 85; midwives, 30; opportunities for employment, 31, 60; professional firsts for, 60, 170, 175n7, 197, 199, 201; recognized, 98, 162; weather observers, 24

women in science, suggested reading, 193–94

Women's History Month, 98

Women's Progressive Club, 61

Wong-Staal, Flossie, 201

woods. *See* forest

World Food Congress, *73*

World Health Organization, 41–42n11, 199

World Seeds, Inc., 75

World War I, 39, 58–59, 60, 157

World War II, xix, 25, 60, 81; post–, 96, 98, 144, 146; pre–, 83, 143, 157; refugee camp, 97
World Wide Web, 202
Wriggins, Howard, 75
Wyoming, 67

x-rays, *32*, 198

Yale University Peabody Museum, 101–2n13

Zybach, Frank, 200

About the Editors and Contributors

Editors

Debra A. Reid, PhD, is curator of agriculture and the environment at The Henry Ford (THF), adjunct professor at the University of Illinois, and professor emerita at Eastern Illinois University. She is author of *Interpreting Agriculture at Museums and Historic Sites* (2017) and coauthor with David D. Vail of *Interpreting the Environment at Museums and Historic Sites* (2019). Other books and articles address rural minority history, including *Reaping a Greater Harvest: African Americans, Rural Reform and the Texas Extension Service* (2007) and *Beyond Forty Acres and a Mule: African American Landowning Families since Reconstruction* (coedited with Evan P. Bennett, 2012). She is a fellow of the Agricultural History Society, was a recipient of the John T. Schlebecker Award from the Association for Living History, Farm and Agricultural Museums, and was the first recipient of the Reid Award for Service from the Midwest Open Air Museums Coordinating Council.

Karen-Beth G. Scholthof, PhD, is professor of plant pathology and microbiology at Texas A&M University. She has published more than one hundred peer-reviewed manuscripts and book chapters on molecular plant virology and host-virus interactions. She coedited *Tobacco Mosaic Virus: One Hundred Years of Contributions to Virology* (1999) and has published more than a dozen articles and book chapters on the history of plant pathology and virology and numerous book reviews. Scholthof has received several awards recognizing her excellence in teaching and research. She is a fellow of the American Phytopathological Society, the American Academy of Microbiology, and the American Association for the Advancement of Science.

David D. Vail, PhD, is associate professor in the Department of History at the University of Nebraska at Kearney. He received his PhD at Kansas State University with training in environmental history, science, and technology. Vail has worked with universities and humanities councils to spread the word about the intersection of policy, science, and agricultural production. His research has appeared in numerous academic journals, such as *Agricultural History, Endeavour, Great Plains Quarterly, Kansas History, Middle West Review,* and *Great Plains Research.* His award-winning book, *Chemical Lands: Pesticides, Aerial Spraying, and Health in North America's Grasslands since 1945* (2018), is part of the University of Alabama Press's NEXUS Series: New Histories of Science, Technology, the Environment, Agriculture, and Medicine. His second book, with coauthor Debra A. Reid, *Interpreting Environment at Museums and Historic Sites,* was published by Rowman & Littlefield in 2019.

Contributors

Kristin L. Ahlberg, PhD, is a historian and assistant to the general editor in the Office of the Historian, US Department of State. She received her PhD in diplomatic history from the University of Nebraska in 2003 and is the author of *Transplanting the Great Society: Lyndon Johnson and Food for Peace* (2008). Her articles appear in *Agricultural History, Diplomatic History, Great Plains Quarterly,* and *The Public Historian.* Ahlberg serves on the Agricultural History Society Executive Council, the National Council on Public History Governance Committee, and the Western History Association Public History Committee.

Matt Anderson, MA, is the curator of transportation at The Henry Ford in Dearborn, Michigan. He oversees the development, care, and interpretation of the museum's collection of nearly three hundred motor vehicles. Matt also oversees the museum's collection of horse-drawn carriages, railroad locomotives and rolling stock, and aircraft. Matt currently serves on the boards of the National Association of Automobile Museums and the Society of Automotive Historians. He has previously worked at the B&O Railroad Museum in Baltimore, Maryland; the Studebaker National Museum in South Bend, Indiana; and the Minnesota Historical Society in St. Paul, Minnesota.

Melanie Armstrong, PhD, is an associate professor in the Haub School of Natural Resources and director of the Ruckelshaus Institute at the University of Wyoming. She studies how social systems are built around shifting ideologies of nature, drawing theory and methods from fields of political ecology, environmental history, and science and technology studies. She is the author of *Germ Wars: The Politics of Nature and America's Landscape of Fear* (2017) and coauthor of *Environmental Realism: Challenging Solutions* (2017). Her fifteen-year career with the National Park Service also provided a laboratory for exploring complex natural resources issues in the landscapes of the West.

Benjamin R. Cohen, PhD, is a historian of science, technology, and the environment at Lafayette College in Easton, Pennsylvania. He is the author most recently of *Pure Adulteration* (2019) and coeditor of *Acquired Tastes* (2021) with Anna Zeide and Michael Kideckel. Writ large, his writing explores the historical origins of modern industrial food systems while working regionally with students and community members to pursue more just and equitable food networks for the future.

Brian "Fox" Ellis, BS, is a professional storyteller and science educator and principal of Fox Tales International. He holds a degree in science education and gained experience working as a resident naturalist at summer camps and conducting fieldwork in ecology. He first told folktales with ecological themes and then began portraying John James Audubon during the 1990s. The Field Museum in Chicago invited him to portray Charles Darwin in connection with the bicentennial of Darwin's birth, and he expanded his repertoire to include Meriwether Lewis, Gregor Mendel, Prince Maximilian, and Robert Ridgway. He is the membership and outreach coordinator for Illinois Audubon Society.

Jajuan S. Johnson, PhD, is a scholar of Africana Studies who focuses on race, gender, and class. He is the Mellon Foundation Postdoctoral Fellow at William & Mary's Lemon Project: The Journey of Reconciliation.

Cherisse Jones-Branch, PhD, is professor of History and dean of the Graduate School at Arkansas State University. She is the author of *Crossing the Line: Women and Interracial Activism in South Carolina during and after World War II* and the coeditor of *Arkansas Women: Their Lives and Times*. In 2021 she published *Better Living by Their Own Bootstraps: Black Women's Activism in Rural Arkansas, 1913–1965*. Jones-Branch is currently working on a third book project, titled *"To Make the Farm Bureau Stronger and Better for All the People": African Americans and the American Farm Bureau Federation: 1920–1966*. She is also a fellow and current president of the Agricultural History Society.

Claus Kropp, MA, is head of the Lauresham experimental archaeological open-air laboratory at the UNESCO World Heritage site of Lorsch Abbey. He is Magister Artium in medieval history and pre- and protohistory (University of Heidelberg), coeditor and frequent contributor to *Laureshamensia*, the Forschungsmagazin des Experimentalarchäologischen Freilichtlabors Lauresham (2017, 2019, 2020, 2021), facilitator for international engagement between museum staff and practitioners on Draft Cattle, a guest lecturer at the Heidelberg Center for the Environment, and serves on the board of the International Association of Agricultural Museums, currently as president (2020–).

Bethann Garramon Merkle, MFA, is an associate professor of practice at the University of Wyoming (UW) and directs the UW Science Communication Initiative. She works at the intersections of art-science integration, equity in STEM, organizational change, and scicomm theory and practice. She has trained more than two thousand scientists in inclusive scicomm and serves in leadership and editorial roles for several journals and professional organizations. As a first-generation college student, she was a cultural and linguistic minority immigrant in French Canada. She started with a wide-ranging alt-ac career and now works to demystify the hidden curriculum of academia and make science-allied careers inclusive and equitable.

Robert Oleary, BS, is a former math teacher from Southeast Michigan. He holds a Bachelor of Science in Secondary Education—Mathematics from Eastern Michigan University. After teaching in Ypsilanti for just over four years, he made the transition to the museum space. He has been a program manager on the Learning & Engagement team at The Henry Ford since 2018.

Emily Pawley, PhD, associate professor of history and Walter E. Beach '56 Chair in Sustainability Studies at Dickinson College, teaches and researches the histories of capitalism, science, and the environment. Her book, *The Nature of the Future: Agriculture, Science, and Capitalism in the Antebellum North* (2020), won the History of Science Society's Philip Pauly Prize. She has written on apples, cattle, and moon-farming and is currently working on projects on climate solutions history, climate crisis pedagogy, and incorporating climate-crisis knowledge into living history museums.

Patricia J. Rettig, MLS, is the archivist for the Water Resources Archive at the Colorado State University Libraries. She earned her MLS from the University of Maryland, College Park, and her BA in English from Wittenberg University. She has been with the Water Resources Archive since its beginning in 2001. In 2014 she received the Friends of the South Platte Award.

Aimee Slaughter, PhD, is an independent scholar and public historian with research interests in popular understandings of science and technology and in museum engagements with difficult histories. She earned her PhD in History of Science, Technology, and Medicine from the University of Minnesota in 2013. Her previous work with the Los Alamos Historical Society included curation and educational outreach, and she served as the lead historian on the 2021 interpretive plan for the J. Robert Oppenheimer House. Her contribution to this volume was written on the unceded ancestral lands of Tewa- and Keres-speaking Pueblo peoples.

Roger D. Turner, PhD, is a museum curator at the Science History Institute. His dissertation, titled "Weathering Heights," traced the intertwined development of aviation and meteorology during the first half of the twentieth century. His recent digital exhibits have examined weather workers, critical metals, mechanochemistry, and five scientific "Instruments of Change." He also cultivates a personal collection of rare but not valuable twentieth-century meteorology artifacts, which sometimes feature in stories on Picturing Meteorology.com.

April White, MA, is the director of the G. W. Frank Museum of History & Culture in Kearney, Nebraska, and an adjunct instructor in the Department of History at the University of Nebraska at Kearney (UNK). A native Nebraskan from the Omaha area, she came to Kearney in 2003 to attend UNK. There she earned a Bachelor of Arts in Art History and a Master of Arts in History. As director and instructor, she engages and cultivates relationships with UNK students and faculty to train future museum professionals to build a bright future for her museum and the public-history field.

Sam White, PhD, is professor in the faculty of social sciences at the University of Helsinki and editor of the *Palgrave Handbook of Climate History*. He has authored numerous books and articles on climate and society and currently leads the Past Global Changes—Climate Reconstruction and Impacts from the Archives of Societies working group.